Little House
OFF THE GRID

Little House OFF THE GRID

OUR FAMILY'S JOURNEY TO SELF-SUFFICIENCY

Michelle and Cam Mather

AZTEXT PRESS

Aztext Press
Tamworth, Ontario Canada K0K 3G0
michelle@aztext.com • www.aztext.com

Library and Archives Canada Cataloguing in Publication

Mather, Michelle, 1960-
    Little house off the grid : our family's journey to self-sufficiency / Michelle
and Cam Mather.

Includes bibliographical references.
ISBN 978-0-9810132-5-1

1. Self-reliant living--Ontario. 2. Sustainable living--Ontario.
3. Ecological houses--Ontario. 4. Mather, Michelle, 1960- --Homes
and haunts--Ontario. 5. Mather, Cam, 1959- --Homes and haunts--
Ontario. I. Mather, Cam, 1959- II. Title.

GF78.M37 2011   640   C2011-905149-4

Printed and bound in Canada

This book is dedicated to the best neighbors on the planet,
Ken and Alyce Gorter. You'll read their names numerous times
throughout this text because they have been such an essential
part of helping us to create the fantastic life we enjoy in our little
house off the grid. Ken has displayed infinite patience while
teaching me practical skills like welding and wiring as well as
helping me to pour concrete and put up a wind turbine. Alyce
has helped us to learn the ways of the country, find things of
great value that don't cost anything, provide us with the "big
pets" we've enjoyed hosting in our paddock and made generous
contributions of horse manure for our gardens. There have been
few steps forward here at Sunflower Farm that haven't involved
their help. We love Ken and Alyce and they have become
family, and we hope they'll accept this small recognition of our
appreciation for their help in making this wonderful place in the
woods a livable and independent piece of paradise.

Acknowledgements

We are grateful to the many people who have helped us with this book.

Ellen Horak read through the text and gave us her perspective about her own move to the country.

Deborah Sowery-Quinn took time out of her busy life to proof the text and provide valuable feedback on the first draft.

Our friend and neighbor Heidi Lind was also kind enough to proof the text and find Cam's mathematical errors.

We are grateful for their help and want to acknowledge that Cam and Michelle are responsible for any remaining errors in the text

Cam is grateful to his mother who found this place for us. She has been gone for more than a decade now but left a huge impact by helping us to find this magical place that we call "home." His Dad had the insight decades ago to move to Eastern Ontario, so that when we were ready we knew should live in a place filled with lakes and forests that was still affordable.

Michelle is grateful to her Dad for all of those Sunday drives in the country, which apparently helped to inspire her desire for a rural life and to her mom for sharing her own stories of living in an off-grid cabin during the early years of her marriage.

We are both grateful to our daughters, Nicole and Katie, for being a part of this journey and not complaining ... too much.

As always we thank our partners in Aztext Press, Bill and Lorraine Kemp who help us keep our dream of publishing books about sustainable living alive and save us from having to commute to a real job in the city.

Thanks to everyone that has been part of this story, whether you were mentioned in the book or not.

Table of Contents

1 Introduction

"I've made a horrible mistake!"

As I stood in the blizzard, I was having serious doubts about the wisdom of my vision. It was two o'clock in the morning. I had a moving van backed up against the garage door, but the wind was blowing the snow so hard it was starting to make drifts on the garage floor. This house I was about to move into was 4 kilometers (2 ½ miles) from the nearest human being. If I needed help tonight I wasn't likely to find it.

I had just had the most harrowing drive of my 39 years. Starting out from the calm of a suburb west of Toronto with a large rental truck, it had started to snow lightly just beyond the city limits. It increased in intensity as I drove east along Highway 401 towards Napanee. As it got heavier, it turned into a dizzy, mesmerizing sort of dance... one where it is easy to zone out or fall asleep.

As I started up Highway 41 north, there were no longer dividing lines separating the two lane road, and maneuvering this monstrous truck without guide lines was proving to be a challenge.

But the worst was yet to come. After 30 kilometers (19 miles) if I made it to Tamworth, I still had to drive this truck through rapidly accumulating snow, another 13 kilometers (8 miles) to my new home and I knew that this late at night it wasn't likely the snow plow would have been through. That's the downside to living on a road far from town. Why should they bother to plow until the morning - only an idiot would be out tonight. Plus there was this really steep hill up an escarpment with a 90-degree turn at the top with ponds on both sides. What was I thinking?

Yet by some miracle I was able to steer this beast through the snow. Now I had to start a fire to warm up the house that was well below the freezing point, and unload this truck by morning. Oh sure, a logical person would have rented the truck for two days, but logic seldom ruled my decisions. My strategy had been to pick up the truck at noon, have it loaded by 4pm, get to Tamworth by 8pm, unload the light stuff by 10 pm, have a relaxing sleep, unload the larger stuff with the help of my

A watercolor by local artist Barry Lovegrove of our new home, built in 1888.

dad who was coming early in the morning, and then have the truck back comfortably within the 24 hours allotted.

But now it was 2 am and things had not gone according to plans. The house had taken longer to empty than we'd expected so I'd got away later. Then there was the snow.

And what kept resonating through my brain was the conversation I'd had with my wife Michelle months earlier as we discussed moving in January. "Moving in January is the stupidest idea I've ever heard!" My reply was "Michelle, we haven't had any real snow in Burlington in January for years. We're lucky if we get a dusting. Trust me, no problemo, it's a no-brainer, don't worry, I've got this under control."

Clearly, I hadn't.

And this was turning into a serious mess.

I was completely exhausted. I had been running on nervous energy for days, making the final arrangements and packing up the last of our belongings. I hadn't eaten properly all day, and now at 2 am, with no energy and frozen fingers, I had to unload a truck in a few hours, if I had any hope of sleep, that had taken 4 of us, 4 hours to load. Math is not my strong point, but this equation wasn't adding up very well.

And as the magnitude of the task ahead sunk in, it occurred to me... what the hell was I doing?

I was uprooting my family from a warm, secure, bungalow in a sleepy city where I was minutes away from the customers of my electronic publishing business that supports us. I'm dragging them 3 hours away to the woods of Eastern Ontario, where our nearest neighbors are 4 kilometers (2 ½ miles) to the east and 6 kilometers (4 miles) to the west, to a home with no power lines to the house, no phone lines, no gas lines for heat, a volunteer fire department that would take 30 minutes to get there in a best case scenario, and police service that might make it in 45 minutes... on a good day... with good conditions.

I've done many stupid things in my life, and clearly, this one took the cake.

* * *

That day 13 years ago was a long time ago, but it is seared into my psyche. The sheer terror that this endeavor invoked that night is still vivid in my mind. And like so many of the tectonic changes in our lives, the outcome on my life was monumental.

It could have gone either way. We could have succumbed to any number of the challenges nature and technology threw at us. Customers could have said dealing with someone 3 hours away wasn't working. Our daughters could have rebelled about the distance from the nearest shopping mall. Trying to live with 20% as much electricity as the average North American does, could have proven just too much of a hassle, and we could have reclaimed our place, in a tract house, in a subdivision that had once been farmland, in a home that looked dramatically like our neighbors.

But we didn't. We persevered. And the outcome has been outstanding! I have experienced joy and fulfillment like I never knew in the city. Our cozy little home nestled amongst thousands of acres of woods is a ray of sunlight emanating from a place at peace with its surroundings. It fits in. I fit in.

It hasn't been all smooth sailing. There have been lots of bumps along the road.

But the bumps make it interesting. The lows make the highs higher.

The journey has been worth every minute of it.

From that night of terror in the January blizzard, to the month without reliable contact with the outside world, I have no regrets about our decision to move off the grid. I wake up most mornings, energized with the thought that I don't have to go out my door, and get in my car, and drive to an office, and sit in a cubicle moving widgets on my computer. I can sit in my kitchen, surrounded by green on all four sides in summer, or with warmth radiating from our woodstove in winter, and sunlight hitting my solar panels and making the electricity I'll need for the day.

My energy sources for the day will be finite. My spiritual energy will be limitless for whatever task I'm taking on that day.

I do not have all the creature comforts of my fellow citizens of North America. I do not have a hot tub. I do not own a dryer. I do not own an SUV. I do not take flights on jets to exotic destinations. But I do have peace of mind. I know where my electricity comes from and its effect on the planet. I know where my food comes from and how it was grown. I know what my impact on the planet is, and how it affects my fellow inhabitants of the planet... rich and poor, and I'm comfortable living with less.

All I have to show for it is peace of mind and happiness.

This book is about our journey and why we hope you might consider starting one of your own. Why you should do an inventory of your house, and your bank account, and your soul, and make a plan about how you can start a path towards your own happiness. It's not about the money. It's about your ability to rage against the constant bombardment of a capitalist society that tells you that this next product or service will bring you joy. We both know it won't, but somehow most of us remain trapped on that treadmill, dreaming about that far off day when we can retire to that dream that may remain forever elusive.

My advice is don't wait. Do it now.

And when you find yourself terrified, standing in that garage with the snow blowing in around you, asking yourself what the hell you're doing, you'll know you're finally on the right track.

A Note About the Format of this Book

While this book was very much a cooperative effort, we each tackled the chapters that we were best suited to write. We identify the author at the beginning of each chapter. This may seem unconventional for a book, but as self-employed, home-schooling, vegetarian, off grid living, solar-powered bike riding environmentalists, that's the story of our lives.

2 Sustainable in Suburbia

By Michelle

Our journey to sustainability began many years ago when Cam and I were just university students in Peterborough, Ontario. At that time there were no municipal recycling programs and the only way to responsibly recycle your cans, newspapers and glass bottles was to take them to a volunteer-run recycling center. Back then Cam and I would look at these materials as we threw them into the garbage and question the practice of landfilling such useful material. Couldn't (and shouldn't) these materials be re-used or recycled, rather than filling up landfills? We made the effort to gather up these materials and take them to a recycling depot and continued to do so when we moved to Burlington, Ontario a few years later.

Now, of course, most municipalities offer recycling programs as part of their garbage/waste pick up services. In fact, some municipalities have even begun to recognize the value and potential in organic materials that were being landfilled and they are picking these up separately and are composting on a large scale.

We continued to look for other ways to tread more lightly on the earth. At that time there was a small, volunteer-run organization called "Friends of the Earth". (This organization, which was begun in 1978, still exists and is affiliated with the international group.) We relied on information from groups such as this to make us more aware of the various environmental issues and our own household practices. We also tried to share what we were learning with our friends and family members and so Cam published a small, two-page newsletter which he called "Friends of Friends of the Earth" and invited everyone we knew to join us in our quest for a more sustainable lifestyle.

We joined a local group called "The Conserver Society," which promoted a conserver approach, rather than a consumer one, which is so predominant in our society. We were reminded that the first R in the well-known 3R's is to reduce and as a family we began to thoroughly examine what we were purchasing and we began to make different choices. One example of this was paper towels. Virtually every North American household purchases and uses paper towels for spills, etc. However, paper towels are made from trees and once they are used they become landfill. Did it make sense to destroy a beautiful tree for a product that is used once and then sent to a landfill? Why not use rags that could be made from our old, worn-out bedding and clothing? Rags that could be used, washed and re-used over and over again. Not only were we doing a good thing for the planet but also our family budget benefited from not purchasing paper towels. The same approach was taken regarding many other purchases – instead of paper tissues we use hankies. Many people turn their noses up at this, repeating the old marketing phrase "It's like carrying a cold in your pocket" referring to the germs. However we soon discovered that when you use a hankie during a cold your nose doesn't become red and sore like it does when you use paper tissues. Paper serviettes and napkins were replaced by cloth ones that I washed over and over again. We stopped buying anything of a disposable nature, whether it was disposable razors, paper plates, plastic cups and when we had our daughters we even limited how many paper diapers we used! That was 25 and 23 years ago when cloth diapers were not as trendy as they are now and the only diapers that were available locally (without using a mail-order service) were the flat ones held on with pins. We bought 2 dozen of those flat, cloth diapers, used them for both of our children and we are still using them as rags today!

At that time we owned two vehicles, a Honda Accord and a Toyota Tercel. Both were reasonably fuel-efficient vehicles and we walked and biked as often as we could. When the girls were too young to ride their own bikes, we discovered a bike trailer that was being made by a company in Oregon. It attached to the back wheel of your bike and you could strap two kids into the trailer and go off on bike rides. Problem was, the trailer was about $700, which was a bit out of our price range! (We were surviving on Cam's earnings running his own desktop publishing business…. in other words, we weren't rich!) We had been toying with the idea of giving up one of our cars – after all, it meant cutting our automobile expenses in half, but we weren't sure whether or not we could survive with just one car. So we decided to take the insurance off of one of our cars – just for 6 months, and we used the money that we saved to buy the bike trailer.

We loved the bike trailer and not only were we able to go for long bike rides with the girls safely riding in the trailer, but it meant that we could use the bike for trips to the grocery store and the library and we had all the room in the trailer for hauling things home. After six months we discovered that it hadn't been much of an adjustment to have just one vehicle so we sold our second car which not only lowered our transportation expenses but our carbon footprint as well!

We started gardening on our little city lot too. We had two enormous Black Walnut trees in our backyard and anyone who knows anything about Black Walnut trees will tell you, you can't have a vegetable garden near them! Not only would they provide too much shade for the garden, they actually contain a chemical called "juglone" which inhibits the growth of many plants such as cabbages, eggplants, peppers, potatoes and tomatoes. So we used our front yard for our vegetable garden, and eventually we even grew corn in our front yard. Needless to say, our neighbors and friends often commented on our "interesting" landscape design!

To improve the quality of the horrible clay soil of the area we added the results of our compost bin but discovered that we needed much more compost than we were able to provide. Then we noticed how other people in our neighborhood seemed to be at war with their leaves and would spend many hours on fall days meticulously raking up every leaf off of their property and filling garbage bags full of them! At that time these leaves were being sent to the landfill site and so Cam started taking our bike trailer or wheelbarrow around and loading it up with bags of leaves. He'd bring them home, empty them into a huge pile and let them sit over the winter and spring. When they had broken down enough he'd add them to our gardens and improve the soil for free!

Most of our neighbors didn't have a problem with us making use of their leaves and Cam usually knocked at a door and made sure that the leaves had been put out at the curb for garbage pick up before he took them. One night he knocked at a door and asked the man who answered if it was okay for him to take his leaves. The man became quite angry and indignant and told Cam that he hadn't gone to all of the work raking up his leaves to have Cam take them for free. Apparently he'd rather that his bags of leaves end up in the landfill before he was going to let anyone else benefit from them!

At around this time Cam also started to notice how many of our neighbors were bagging their grass clippings and sending them to the landfill site too! He didn't want to put grass clippings into our compost, partially because many of our neighbors employed lawn-spraying companies and we didn't want to be adding those toxins to our compost pile. However, he recognized that sending these to the landfill was not a sustainable practice either. He ended up appearing before our city council to ask that they enact a bylaw banning grass clippings from the garbage. City council was not too interested in his idea until he looked at the statistics on garbage and discovered that the volume (and weight) of garbage increased quite dramatically during lawn cutting months and he was able to show the city council that they would be saving $200,000 by banning grass clippings from the garbage. This was a time when a conservative provincial government had reduced how much money they sent to municipalities for things like waste disposal. It was unfortunate that it took an argument about "eco-nomics," rather than "eco-logic" to convince them, but in the end grass clippings were banned! He basically made them an offer they couldn't refuse.

We acquired a new "pet" around this time as we continually added to our compost piles. One day the woman who lived directly behind us showed up on my front porch. In the five years we'd been living in the house, this woman had never spoken to me before and in fact had never even waved back when I had tried to be friendly to her! Now she stood on my front porch and tried to make small talk. Soon the real reason for her visit became clear. She had noticed a rat scurrying along the fence that separated our back yards. Her dogs were being driven crazy, barking at this rat as it went about its business.

As much as I wasn't thrilled to have a rat in the backyard (something tells me that he wasn't the only one) I wasn't about to use dangerous poisons to kill it either. I had a "live and let live" attitude even then and as long as the rat wasn't trying to share my home with me, I decided I would let him be. The fact that this woman who had snubbed every effort I had made to be neighborly was suddenly asking for my help didn't sit well with me either (petty, I know!). So I assured her that we were well aware of "Rodney" as my daughters had christened him and that he wasn't causing us any problems. She left in a huff and "Rodney the fun rat" carried on with his life in our backyard!

Cam started to realize that to make the biggest impact on our city's practices it would be useful to join a committee and when a vacancy came up on the Sustainable Development Committee he applied and was accepted as a new member. He was involved in many projects and events during that time but the most memorable one for our family was a video that the committee produced called "Changes: A Low-Impact Environmental Lifestyle". The idea for the video came when Cam was lamenting the fact that every year in Burlington (as well as other cities) there were numerous house tours during the holiday season in which big, beautifully decorated homes were open to the public to tour, usually with all proceeds going to a charity. Cam thought it would be great to offer "green" tours of a sustainable home instead of the unsustainable "monster" homes that were on display. He broached the subject with me, wondering if I would be game to open up our home as an example of a more sustainable one. I wasn't too fond of the idea, especially since we had the opposite of a monster home – a two-bedroom, less than 1,000 square foot bungalow and I didn't relish the thought of crowds of people arriving to tour it!

Instead, the Sustainable Development Committee came up with the idea of putting a tour like this on video. They encouraged the participation of local high school students and spent some time at our house filming us and we showed the various ways in which we attempted to live more sustainably. The video was a huge success and was sent all over the world as an example to other cities!

Another practice that was to become an annual event for us began around this time. We began to become more and more aware of the deforestation of not only our country but the world and the more we read the more we realized that someone needed to be planting trees, rather than just constantly cutting them down. The problem was, our 40' x 108' city lot already had as many trees on it as it could manage. So we began, through our participation in the Conserver Society and other groups, to plant trees all over the place – in Bronte Creek Provincial Park, in the local schoolyard, etc. This is a habit that we have carried on, even while living here on 150 acres of trees. Every year we order seedlings from a tree nursery and plant some more trees on this property. I'm sure our neighbors around here question our sanity... after all, don't we have enough trees here in the forest?

Not surprisingly some of the events and causes that we were involved with in the city caught the attention of the local newspapers. Not only did we have full-page lifestyle stories written about us, but Cam or I were often called by local reporters to get our impressions or opinions regarding local news of an environmental nature. We both learned pretty quickly to treat these interviews very seriously, as on off-hand remark made in jest can turn out quite differently when you are quoted in the newspaper. My favourite example of this was when I discussed our vegetarianism with a reporter. I joked that when my daughters became teenagers I wasn't worried about them experimenting with drugs but rather I figured that they they might experiment with meat. This comment, which I only meant as a joke, made it into the news stories, while many of the very serious and important comments that I had made during the interview did not!

Eventually we were able to get our garbage production down to about 4 cans of garbage a year, for our family of four. Every year during "Waste Reduction Week" there was a steady stream of local TV News crews and newspaper reporters calling and visiting to find out how we did it.

Apparently the same local reporters that had depended on Cam for his "eco" comments throughout the years were going to miss having him to rely on. Just before we moved away from our suburban home the local paper ran the headline "Mr. Environment Moving – For Good"! Our adventure was about to begin.

THE HAMILTON SPECTATOR Friday, January 8, 1999

Mr. Environment moving — for good

Mathers escaping from urban life

By CARMELA FRAGOMENI
The Spectator

Cam Mather gave his family the ultimate Christmas gift, and a healthy fresh start for the new year.

Burlington's Mr. Environment and Conserver is leaving urbanism behind and moving to a solar-powered farm on 61 hectares (150 acres) in eastern Ontario at the end of this month.

"We're dropping out," he says happily. Dropping out in favour of fresh air and peace and quiet.

Mather says he and his wife Michelle have been able to buy the farm, northeast of Kingston, because they got off the consumer merry-go-round.

> "I'm just so turned off by a culture that spends so much time consuming at the mall when people already have what they need."
>
> Cam Mather

"We stopped buying mindless stuff and paid off our mortgage. We're just in our 40s and we can now move out to the country, and the fresh air."

He says Christmas, for environmentalists like him, is an oxymoron. They want to celebrate it, but so much is focused on consumption, it has become an environmental nightmare.

"I'm just so turned off by a culture that spends so much time consuming at the mall when people already have what they need. There seems to be this silly obligation that we need to buy something for everyone."

The Mathers don't buy gifts at Christmas because of their strong commitment to cutting

Burlington environmentalist Cam Mather is trading in his home — and compost heap — for life on his farm near Kingston.
Spectator file photo

3 And We Finally Found What We'd Been Looking For...

By Michelle

I come by my desire to live in the country quite honestly. My dad was born and raised on a farm north of Barrie, Ontario. As one of 13, he did not inherit the family farm and ended up working at a steel mill at Stelco in Hamilton after his time overseas in WWII. He and my mother raised eight of us, first in Hamilton and later in the suburbs of Burlington, Ontario. I fondly remember many "Sunday drives" taken through the countryside of north Burlington as my father dreamed of buying a piece of land and returning to his rural roots. Somehow it seems that his dream rubbed off on me, and as an adult living in suburbia I looked longingly at country places and dreamed of one day living on my own little piece of paradise.

Our search took about five years, off and on. During the "on" times we would scour newspaper ads, magazine ads and watch for realtor signs during our drives. There was a handy property listing from a company called Dignam (www.dignam.com) that featured pages of acres of bush throughout Ontario. Time and time again we would become discouraged by the price of real estate and our inability to find just the right place. Friends of ours took a trip to the east coast of Canada and we asked them to check out a property that we'd heard about while they were down there. They returned with an envelope full of photos but even without seeing the place with our own eyes, we knew it just wasn't "quite right." Besides, even though the lower real estate prices were more within our price range, we really didn't want to have to go that far away from our comfortable surroundings.

That was just about the time that the Canadian Province of Quebec had a referendum to separate from Canada and form its own country. That night as the results flowed in, commentators theorized that if Quebec did indeed leave Canada, the Canadian Maritime provinces might decide to align themselves with the United States. We love our universal health care, and moving to the Maritimes and losing it didn't seem like a good proposition.

At some point we decided it was time to put together a "Wish List" – an inventory of the most important characteristics of our "dream place." This really helped us to sort out between the two of us what we were really looking for – what was most important – and which things were more than likely "pipe dreams." Here's what our Wish List looked like:

Property Requirements / "Wish List"

- It must be private and out of earshot of any neighbors
- It should be 15 to 30 minutes from the nearest town
- Property should have 50 to 100 acres, at least
- A good phone line or access to one will be very important for our business
- Access to water would be an asset, be it pond, stream or lake
- We would like 4 bedrooms and hopefully a separate living room, dining room and family room
- At least half of the acreage should be forest – sugar maples would be great
- An area for a large garden, preferable with soil that has been worked previously
- Hopefully the garden could be certified organic
- We would be interested in a house on the property but we are not averse to building our own home
- If there is a home, any form of passive solar heat or solar power would be a definite plus
- A small barn or outbuilding is not necessary, but would be useful
- We do not want deeded right of ways or other encumbrances to our property
- We require mineral and timber rights to our property
- Price range $110,000 to $150,000

We used this Wish List in a number of different ways. We sent it out to realtors in areas that we liked and asked if they had any listings that were along the lines of what we were looking for. When we responded to an advertisement, we used our wish list to ask questions about the listing. This technique didn't always work however, as we found that realtors were occasionally so determined to get us to come and see one of their listings that they would somehow overlook one of our significant "wishes".

For example, we spoke to a real estate agent in Eastern Ontario about a listing and faxed through our wish list. The agent insisted that the property she wanted to show us was perfect for us. We drove 4 hours to the agent's office and then rode with her as she took us to the listing that she described as a "perfect match" for our wish list. As we drove towards the property, Cam and I even began to question her, item by item. The first item was that our dream property "must be private and out of ear-shot of any neighbors." Even as we drove towards the property she insisted that this listing was all of that and more! As we turned into the driveway of the property that she had listed, Cam and I were both dismayed to see that the house was directly across the road from another home. This home had neighbors that were so close you could whisper and hear each other! When we pointed this out to the agent she said, "Oh, I forgot all about that house. Oh well, they are an elderly couple, very quiet, so they won't be a problem ..." Right away Cam and I both began to envision what would happen when the lovely, quiet, retired couple decided to move away and sold their home to folks with noisy ATVs, snowmobiles, or boom boxes, or better yet, a biker gang! After living in a "fishbowl" in suburbia, we were determined to find some property with some privacy!

I lost count of how many places we looked at. We soon found that our desire to live in an old farmhouse full of charm and character was at odds with our desire to have a home that was open and sun-filled. Most of the century-old farmhouses that we looked at were dark and dreary inside, divided up into many little rooms and not only were they not suitable for passive solar heating, they would have even been difficult to heat with a woodstove since the rooms were so closed up and segmented. When we did find old homes that had been renovated and upgraded they were often beautiful but way out of our price range. The ones that we could afford were the ones that had not been upgraded and not being handy types ourselves, we were not thrilled with the prospect of buying a "fixer upper"!

It was amazing to see some of the homes that folks were offering for sale – one could only be accessed across a neighbor's property, one had an enormous hole dug into the ground next to it where they had been "thinking" of putting in a pool, one house had a curvy set of stairs that were also on a slant – even as a young person I couldn't imagine going up and down them very often – I hated to think of doing it as I got older!

Our search became an off and on again thing and would definitely be more "off" than "on" whenever we had been disappointed yet again about what had seemed to be a promising lead. It was during one of our "off" times that Cam's mom called him and said "I think I've found your house." No doubt Cam didn't get too excited – his mom had found numerous places for us to look at during our search – but he took down the details for the real estate agent and promised to call and get more information. Cam's mom had been visiting a friend in Kingston and had happened to pick up one of those glossy real estate brochures that you see from time to time and had come upon a listing that she thought sounded like it suited us to a "T". She read out the description to Cam;

"Retreat Property". Own power, solar, wind & DC. 3BR, bath, all re-done. Studio over garage, horse barn. Very private 150 acres, woods, ponds $126,900"

Although we didn't see the photo until later, she told him that it showed a typical looking storey-and-a-half farmhouse, white clapboard with green trim.

Cam called the real estate agent and asked all of the usual questions. Finally he told the agent that he would be faxing a "Wish List" and asked the agent to go through the list and tick off the items that were appropriate to this house. We waited with baited breath for the fax to be returned and when it was we realized that he had ticked off every single item on that list! That had never happened for us before and the real estate agent later admitted it had never happened for him either! After another telephone conversation with the agent, Cam decided that we should go and see this place and so we arranged to meet with the agent a few days later.

On a beautiful May day we made the 3-1/2 hour drive from our home to view the property. We left our daughters, then 10 and 12 years of age, at home with family as we felt that there was no sense dragging them around the countryside, possibly getting excited about each property and then being disappointed as yet again it was not a match.

Our drive to Tamworth was a bit terrifying. This was the first place to match all our requirements. What if it really was the spot we wanted? Could we actually leave the safety and comfort of our suburban life to live in the woods? What if it was perfect but we just didn't have the gumption to do it? What if we were all talk and no action?

We were somewhat familiar with the area since it was on our way to Cam's family cottage, which is now an hour north of us. Interestingly enough, acquaintances of ours from the city had moved to this very community two years previously, although we'd never had the opportunity to visit them. Their regular emails describing their new life in the country had been entertaining and educational, especially since they had moved to an off-grid home!

We arrived in the community a few minutes before our appointed time and our agent asked if we could wait for him until he was ready to leave his office. Cam and I used the opportunity to take a walk along the main street and noticed that there seemed to be a number of healthy businesses in town; a bank, a hardware store, a liquor store, some craft shops, a pizzeria, a tearoom and a bakery, among others. The bakery appealed to us after our long drive so we ended up in there. When I saw "samosas" on their lunch menu I asked "are they vegetarian?" I was amazed when the answer was "Not only are they vegetarian, but they are vegan." Wow! (As a vegetarian I was used to having to explain what I meant by "vegetarian" and so to find someone that knew what "vegan" meant was quite surprising!)

After a quick snack at the bakery, we returned to the real estate agent's office and were driven out to the property. The drive seemed quite long that day as we negotiated up hills and around twists and turns and admired one pond after another, mixed forests and big rock outcroppings and a huge Osprey nest in a pond just around the corner from the house. In reality, it is only a 10-minute drive from town, but it seemed much longer that day.

We pulled into the driveway and immediately noticed a rather tired looking old home as well as a much newer building. The grass hadn't been cut in quite some time so it was rather high and unkempt looking. The roof was the worst part of the exterior of the house. Even though the real estate agent had warned us and insisted it would be painted before it sold, it looked pretty bad. Made of metal, which is common in this part of Ontario, it had been painted a dark forest green, but the paint was faded and flaking off, leaving large rust-colored patches all over the roof. Ugh. I must admit that at that point I didn't hold out much hope for the inside of the house and I began to think we had wasted another day to come and look at this place!

However, once the real estate agent let us inside, I changed my mind. Hardwood floors gleamed and light flooded into windows on every side of the house. As we entered the house I noticed a set of stairs going up in the middle of the ground floor. To one side of the stairs was a large living room with a woodstove at the end. On the other side of the stairs was a large dining room with a kitchen at the end and a bathroom next to the kitchen. The living room was completely empty, other than cotton curtains hanging on the windows. The dining room still had an antique table and chairs and an antique cookstove. The bathroom had a claw foot tub and an old pedestal sink. Wow! I liked what I saw inside!

We headed upstairs and found a large, open room above the living room. There was a small bedroom and a very large bedroom, as well as a large closet. Again there was a bit of furniture upstairs, but it was empty enough to see the beautiful hardwood floors in the open area and the Berber carpeting in the large bedroom. I was smitten!

I wasn't quite so impressed with the basement that was cool and dark and empty but I didn't imagine I would be spending much time down there, so I wasn't too concerned.

At that point the agent suggested that we should go and take a look at the other building. This was a two-storey building with a garage on the bottom. A door led into a room with a woodstove and there was also a battery room downstairs with a huge bank (50 at that time) of batteries and various equipment on the walls that I certainly didn't recognize or understand at that time.

A set of stairs took us up to the second floor. Unlike the ground floor, which was cement, the upstairs was all wood – and I mean ALL wood. The floors, some of the walls and the ceilings! There were 4 rooms upstairs – one large room directly above the garage, 2 smaller rooms and a roughed-in bathroom. This building had obviously been built fairly recently – it still had that "new house" smell to it! (I may have been imagining that.) Again this building overwhelmed me – I liked it!

At this point we wandered around the property a bit – the agent showed us the barn, the paddock, the solar panels, the wind generator, etc. I don't remember much of our conversation with him – I'm sure that we asked a lot of questions – but I seem to remember feeling a little dazed. We wanted to take a hike around the property and so our agent suggested that he would leave us but we were welcome to wander around. Cam's parents had arrived while we were looking through the house and so our agent left us to discuss it with them. They, too, seemed quite excited by the house and we discussed the various strengths and weaknesses.

We said goodbye to Cam's parents and went for a short stroll along an old laneway. It was a beautiful spring day and the ground was covered with multi-colored wildflowers pushing up through the blanket of dried leaves on the ground. One of us commented to the other that it felt like walking through a conservation area in the city, but without the crowds.

As we walked we kept saying to each other "Can you imagine OWNING a place like this?"

We returned to Cam's parents and agreed to discuss our observations with them that night at dinner. We had arranged to stay overnight at their cottage with them and so during the whole ride there, Cam and I discussed what we had just seen over and over again. I soon had to pull out a piece of paper to take notes!

Up at the cottage we reviewed the house viewing once again. Cam's parents had made their own list of concerns and questions and the next morning we left to head back to the house to talk to the agent again and get some of our questions answered.

Once again we ended up at the bakery and this time we asked if anyone knew where our acquaintances from the city lived. We were given directions that took us about 20 minutes north of town. We dropped in without notice and once we reminded them who we were and explained what we doing there, we were welcomed with a cup of tea and lots of stories and advice about the town, the area and most importantly about living off the grid!

They were a wealth of information but the one piece of advice that they gave us that day and that has stuck in my mind ever since is "Don't make our mistake ... don't ever move into an off-grid house in November!" November is the gloomiest month in this part of the world – the hours of daylight are getting fewer and fewer and what daylight there is tends to be cloudy and rainy – not exactly prime conditions for generating electricity with solar panels! They told us a horror story of moving into their off-grid home in November and wondering why they kept running their batteries down low. They had to keep running their generator to keep up with their demand and they started to wonder if this was the reality of living off grid.

After a very educational and rewarding visit with them we headed off with promises to keep them up-to-date about our decision.

At that point we headed back to suburbia to consider this house. Cam and I became immersed once again in our city lives while at the same time we kept talking about the property we had seen. At some point we sat down and made a "Pros and Cons" list. Any time that Cam and I have had a big decision to make, we've found that it helps to actually list the good and the bad, the advantages and the disadvantages, and then determine which side is lengthier. This method isn't always as "scientific" as it should be as you can easily skew the results by listing even the most inconsequential items on the "pro" side and conveniently forgetting some of the "cons" but I've always felt that this skewing is part of the process. If I skew the results one way or the other, either consciously or subconsciously, it must mean that deep down I've made my decision!

Needless to say that even after listing all of the advantages and disadvantages of this house, we were able to convince ourselves that there were more Pros than Cons to this decision!

Two weeks after viewing the property we were still waffling on our decision. Then the real estate agent called to tell us that he was going to be presenting an offer on the house to the sellers and thought that we might want to know that, if we were at all serious. Hmmm... this sounded a bit like a realtor's trick. It was a trick that worked and when it suddenly appeared that someone might purchase the house before us, we realized how much we did want it!

So we quickly put together an offer with the help of a lawyer. Our offer to purchase specified the price, what we expected to be included, provided for an inspection by our choice of Home Inspector and also included something that very few people seem to think about; we wanted all Timber and Mineral Rights for our property. This was something that we had read about in a book called "*Heading Home: On starting a life in the country*" by Lawrence Scanlan. Apparently a lot of folks who own land in the country neglect to insist on these and find themselves with logging trucks using their driveway and blasting going on in their back fields for mines! We actually attended a meeting where a person spoke of a company that had opened a limestone quarry on his property near his house. What a nightmare that would be!

We worked out all of the little details with our lawyer, presented our offer (via fax) to the real estate agent and then waited to hear back. I don't remember how long it took to hear the results of our offer but it couldn't have been too long or I think we would have gone crazy! Eventually the agent called to say that our offer had been "chosen" by the sellers and without too much negotiation we came to an agreement.

A few days later we called our friends who were living in our soon to be "new" home village. When I tried to tell them the news she said, "Yes, we know you bought the place and we already know that the closing date is June 19th." I said, "Wow, how do you know all of that already?" As she reminded me, "Michelle, you are moving into a small town and everyone is going to know your business. Get used to it."

4 Planning, Preparing and Packing

By Michelle

Once our offer had been accepted and the closing date had been arranged the real fun began. First we arranged financing for the new place, by arranging for a loan based on the equity in our existing house. Since we had managed to pay off our mortgage a few years earlier, it put us in a good position to borrow some money and not have to worry about selling our existing house quickly. We wanted to take possession of our new place and move into it gradually, making sure to educate ourselves about not only the electrical system but to figure out how to have phone, fax and internet service as well. We knew that the previous owners had managed all of these at the house and so we weren't too concerned about getting them all up and running. These methods of communication are important when you are just living in a house, but we planned on running our desktop publishing business from our new home as well and communication would be very important! Cam will provide more detail about how we managed to set these systems up in a later chapter.

We were quite excited about our move and there were a number of interesting coincidences which made us feel that at some level we were meant to be in our new house. James Redfield's book *"The Celestine Prophecy"* suggests that people need to pay more attention to coincidences in their lives. They have great meaning and we should become more aware of them. Purchasing our place in the country consisted of one coincidence after another.

Several months prior, we had been given a book called *"Heading Home; On starting a life in the country"* by Lawrence Scanlon. It documented his move from the City of Kingston to the small village of Camden East, where he was an editor at *Harrowsmith Magazine*. His parents lived just north of Camden East, in the village of Tamworth where we ultimately ended up. In his book he recommended a book called *"In The Skin of a Lion"* by Micheal Ondaatje, author of *"The English Patient."* The book weaves two wonderful themes of building the Bloor Viaduct in Toronto, and logging in a series of lakes called the "Depot Lakes," which are located near our property. Ondaatje even mentions the Tamworth Hotel in his book. We finished reading this at just about the same time we put the offer in our house, which seemed very coincidental.

After our visit to look at this house, Cam went down to our basement where we stored our back issues of *Harrowsmith Magazine* (later called *"Harrowsmith Country Life"* and no longer published) and came up with a copy a number of years old, which featured the house. He remembered having read it and being intrigued by the idea of being "off the grid". It included a photo of the previous owners Jean and Gary sitting in the living room using a computer.

Shortly after purchasing the new house, one of our neighbors in Burlington called and said to Cam "I hear you've bought some property in Eastern Ontario. Where is it?" Cam mentioned the nearest village, Tamworth, and our neighbor said "Where exactly?" and Cam told him "13 kilometers east of Tamworth". He replied, "I'm coming right over." Cam was intrigued by his behaviour until he arrived on our doorstep carrying a framed black and white photo of our new house. Cam was surprised and asked, "Where did you get that photo of my new house?"

Our neighbor explained that his father had been born and raised in this house and, in fact, the photograph showed a couple standing in front of our new house holding a baby. It turns out that the baby was our neighbor's father! We were all pretty amazed that in the Greater Toronto Area where there are more than 5 million people, we lived three doors down from a guy whose dad grew up in the house 3 hours away that we just bought. He invited us to visit him in a week's time when his parents were going to be there. He asked us to bring photos of our new home to show his parents. When we met his parents the following week and announced that we'd bought some property in Eastern Ontario, his parents were just as shocked and surprised as we had been when we began to pass around photos of "our" new home! It also turned out that our neighbor's mother had been a boarder at our house when she taught school in the one-room schoolhouse nearby and had then ended up marrying the young man whose home she was boarding at. So both of his parents had a connection to our house and we have enjoyed having them visit us at least once a year since we've been here!

Another coincidence occurred when we came up in June to sign the papers at the lawyer's office and to finally meet Jean, one of the previous owners. We liked Jean from the moment we met her and we were glad that she was able to come and show us the electrical system and explain how it all worked. She was also kind enough to suggest that she would take us to meet our new neighbors, Ken and Alyce. We hopped in the car and drove 4 kilometers (2-1/2 miles) to meet our "nearest" neighbors who would end up becoming such a huge part of our lives here. We said our hellos to them and they enquired as to where we were from. When we said "Burlington" Ken immediately asked what high school we had attended. When we told him that both Cam and I had attended "M. M. Robinson High School," Ken replied, "So did I!" What were the odds of that?

There was no question in our mind that we were meant to end up here at the place we now call Sunflower Farm.

Despite having had Jean explain our electrical system to us, it still managed to completely confound us on more than one occasion. After having taken possession of our new house in June, we spent most weekends and a couple of weeks here that summer. On one of the weekends we were busy cleaning – washing floors and windows and just freshening the place up. As I repeatedly freshened the water in my cleaning bucket on the Saturday, the tap suddenly went dry. Oh oh... my first thought was of our well. You often hear of country wells going dry during the summer months. I called Cam and together we investigated. He went out to our electrical/battery room that is located in our second building and noticed that our inverter (the piece of equipment that changes the DC power from the batteries into AC power that is used by appliances and water pumps) had turned off. Hmmmm…. that didn't make sense. It was a brilliantly sunny day and the batteries would have been getting a very strong charge! We had been away all week, so they would have had all week to charge. So suddenly it looked like we were about to move our family to a house that ran out of power in the middle of a brilliantly sunny June day. It was a Saturday, so Cam wasn't able to get anyone on the tech support line at Trace, the inverter company. We'd have to wait until Monday to figure out what had gone wrong. Then, late in the day, once the sun went down, the inverter mysteriously started up again. Whew! Dodged a bullet.

Then on Sunday, the exact same thing happened. Even though it was sunny, the inverter shut off just after lunch. This did not bode well. Maybe the whole solar power thing wasn't such a good idea after all.

When we were finally able to question Jean, it turned out that what was happening was that the batteries had reached their full charge and when the pump turned on, the inverter protected the batteries by shutting everything down. Once the sun went down and the voltage in the batteries dropped as a result, the inverter came back on all on its own.

(I am happy to report that now that we have improved our system, this does not happen any more!) At the time all we would have had to do to fix this was to have turned on some of the "Direct Current DC" lights in the guesthouse which ran directly off the battery bank. This would have drawn down the voltage on the batteries to a level safe for the inverter to start, which would have in turned charged the pressure tanks and got our water flowing again.

During the summer and fall of that year, Cam was successful at first getting a phone working, then a fax machine and finally we had internet. It was time to put our suburban home on the market and make a clean break from the city. We listed our house and worked hard at keeping it clean and tidy, ready for showings at the drop of a hat. Our house sold quite quickly and the new buyer asked for a closing date earlier than we had been planning on. Cam's attitude at that point was "What are we waiting for?" and we discussed the idea of moving at the end of January.

My argument against that plan was that January was a miserable month for a move. Who wanted to move boxes and furniture through snow or slush-covered sidewalks and then drive through snowstorms to get to our new house? Cam argued that this part of Southern Ontario hadn't seen much snow in recent winters and that this winter probably wouldn't be any different. He finally convinced me and our moving date was set for January 28.

In order to avoid the expense of a big moving truck, we rented a small truck a number of times during the late fall and began moving our furniture and belongings slowly over that time. By the end of January, all that was left to move was the furniture and things that we were still using. It was a good thing that we had done a lot of work beforehand, as that January we had 17 days straight of snowfall. The snow banks along the roadways and sidewalks were higher than they had been in years and as we loaded furniture and boxes we had to lift them high over the snow banks. Cam had one particularly harrowing drive here with a loaded truck, which he described in the introduction. Take my word for it – January is

A photo of the house taken from a plane by our neighbor Ken Gorter showing our little house surrounded by lakes, ponds and forest.

NOT a good month to move in this part of the world!

We survived our move and we were welcomed to our life in the country the very night we officially moved in, by being invited for dinner at the home of our new neighbors, Ken & Alyce. Even though I had told my Dad that we were anxious to move out of the city and get away from neighbors, I quickly learned that your neighbors in the country become an important part of your life!

The sign our neighbor Alyce hung on our gate the day we moved into our new home. (Please note, I kept the sign and staged this photo by photographing it in the summer. When we moved in on February 1st we would not have had grass and green leaves on the trees!)

5 E.T. Phone Home – Communication Off The Grid

by Cam

"Mr. Watson – come here – I want to see you."
Alexander Graham Bell's notebook entry of 10 March 1876 describing the world's first telephone call to his assistant, Thomas A. Watson, in the next room

Living in the modern world, we've come to take many luxuries for granted. When it's cold outside, we turn up the thermostat and our living space warms up. When it gets dark, we flip a switch and electricity instantly creates light. When we want some warm food, we turn a dial and it's heated. And when we want to speak with someone in a distant place, we pick up an oddly shaped piece of plastic, hold it to our face, one part over our ear and one part over our mouth, we push some buttons, and miraculously we can speak with someone across town, or across the world. It's easy to take the telephone for granted. Hundreds of millions of inhabitants of the planet have never experienced a phone conversation, but for most people reading this book it's always been there. It's just part of the world, like trees, and houses.

When you live off the grid in a remote location, you do not take the telephone for granted. In fact, it is often a key element in any decision-making about moving off the grid. Many people who are really "hard-core" off-gridders, will forgo this luxury, and decide to live in complete isolation. Their attitude is that they're moving to get away from the trappings of modern society, so they're not being true to the cause if they've got a phone. I can hear them now, "You're selling out man, you've got a phone man, that is such a bummer dude. Come on man, stay the course, go phoneless!"

While I understand and sometimes admire this level of commitment, it was not going to be right for us. We still had enough financial obligations that we were going to have to generate an income, and that was going to be extremely difficult without this magic box that plugged us into the rest of the world. We also had children, and though we had been blessed with very healthy children, when there is someone else in your care the concept of being able to phone for help is important. Also, we heat with wood so there were fire concerns. Regardless of how far away the fire department was, I wanted to be able to call if I needed it. The need for a phone was a given.

The thought of having to walk 4 miles to use the neighbor's phone, or of having a phone at the end of a long country driveway (causing you to carry on your conversations exposed to the elements), left me cold.

Communication off the grid can be a challenge. There are some options:

- Cellular
- High speed wireless, VOIP
- Just a phone line
- Point to Point communication with an Optaphone type unit
- Two tin cans with a long string running between them, strung to the neighbors

Today the easiest option is cellular, if you have cellular coverage, and that's the big "IF." Companies that offer cellular service want to make a profit. To do this they have to locate cellular towers in the areas where they will get the maximum benefit. Installing and maintaining a tower is expensive, so it doesn't make economic sense to put one where very few people would ever make use of it, and therein lays the problem with relying on this for your phone service. If you are far enough away from the built up world, it's unlikely there will be cell service. If there is, great, go for it, it will likely be the most economical service available to you. As more people use cell phones, the cost has dropped. It will still cost much more than a traditional phone, and you often won't have access to as many discount programs and long distance discounts. Once you're a cellular customer, you are pretty much stuck with that service provider and the programs and discounts they offer.

Cellular was not an option for us when we first moved here because the service in our area was marginal at best. When we purchased Sunflower Farm, we purchased the "Optaphone" system from Jean and Gary, the previous owners. This was an additional $5,000 over and above the purchase price of the house, but was the best option for a phone. Jean had researched it thoroughly. The Optaphone is a very high tech piece of technology that takes the electrical impulses that run through a phone line, and converts them to radio frequencies. You have one box that sits at the end of the phone line and one box that sits in your house. Even though you are using radio frequencies, you don't have the problems with the previous radiophones, which were more like CB Radios. In those cases, you had one frequency that the two people talking shared. This meant you would talk, and then say "Over," which let the person you were talking to know that you were finished your sentence and that it was their time to talk. Obviously conversations would be less than spontaneous.

The point-to-point technology in the Optaphone got around this. We actually had two radio frequencies, one for sending and one for receiving. Plus the Optaphone made all of our house phones work just like regular phones, with dial tones and quality similar to a regular phone. And if we wanted to, when we were talking to someone, we could interrupt him or her, because the technology allowed it. The two radio frequencies we used cost us several hundred dollars a year for our license from the Government of Canada, because in Canada radio frequencies are considered publicly owned. The frequency we used was in a range, which was set aside for this type of communication, and no one else would be able to use. After having worked at a radio station in the 1980's I kept threatening Michelle to start broadcasting a pirate radio program on our system, but as would so often be the case with my ramblings, no one would be listening.

On the day of the closing of our house deal, Jean was kind enough to agree to meet us and walk us through the various systems like the phone and solar equipment. Learning about the phone involved a visit to the Feed Mill in Tamworth where the send unit was located. Jean had picked the feed mill because it had a very high tower where she could locate the antenna. Using a point-to-point system involves trying to get your antennas above obstructions so you have a "line-of-sight" so that the antennas can talk to each other. The send unit was located in an explosion proof box downstairs. Feed mills have to be very careful with electronic equipment because the dust can become explosive, and a spark could set one off. We then walked up stairs to the second floor and followed the wire, which ran to the antenna. At the top of the stairs we stepped into this cavernous room with large pipes running through it from the ceiling, which were used to mix the various grain combinations for animal feeds.

As I scanned around the room I saw an aluminum extension ladder fastened to the one wall, extending to a trap door in the ceiling. My first reaction to it was dread, imagining the poor person that had to climb that thing. It had to be 3 stories high, running straight up. It wasn't like a ladder leaned up against your house on an angle, this sucker was vertical. As a child I was fearless of heights and I still cringe at my memory of some of the stuff that I did on roofs of houses that were being built in my subdivision. But that fearlessness has long worn off, and now I am very uncomfortable with heights.

As luck would have it, the first words out of Jean's mouth were "Hope you don't mind heights!" as she scampered up the ladder. So as acrophobic as I was, I also was a man, and admitting to something as unmanly as being afraid of heights was not within my frame of reference. If Jean was going up the ladder, so was I. Now on a good day, a three-story high vertical aluminum ladder would be a challenge for me. But this ladder had something extra to add to the terror. The ladder was covered with this thick, fine, grain dust. You've seen professional bowlers and pool players put dust on their hands so things slip through them. Dust makes things very slippery. Sometimes this is a good thing. This dust was not going to make climbing this ladder any easier. The first two steps were all right, but after that the terror set in, and got progressively worse with each rung I moved up. It was bad enough that the steps of the ladder were slippery but of course the rails I was holding on to were just as slippery. You can tell when you're committed to move to the woods when you'll do something that goes so completely against your nature, because you know without the phone, it isn't going to happen.

Climbing through the trap door at the top didn't offer any comfort either, because the platform where the antenna was located was not level, and the guardrail looked as if a sparrow bumping into it would send it careening to the rocks on the Salmon River below. And the grain dust - the grain dust was there too just to make things fun. I'm sure Jean discussed the antenna with me that day, but I do not recall anything she said. Much like the feeling many people have after meeting a celebrity they're enamored with, but are unable to recall what they talked about.

I had one thing on my mind and it was how I was going to get back down that slippery ladder from hell. Going up it's much easier to not look down. Getting started on the ladder, I had no choice but to look down. I'm convinced the arthritis that I'm starting to feel occasionally in my hands, started that day, as I was actually able to crumple and bend the aluminum rungs with the force I was hanging onto them. Well, all right, I exaggerate, but I was holding on really tight. Once down I could relax and have a normal conversation with Jean.

On the drive back to the house Jean pointed out to us where the phone line ended. She made a rather prophetic comment in passing the importance of which I didn't realize at the time. She said "I always wanted to put a remote box here at the end of the phone line powered by a solar panel so I didn't have to use the mill." This became important as I tried to negotiate restarting the Optaphone with the mill. I would have to get the phone company to hook it up, and arrange a fee to cover electricity expenses for the mill. I did not meet the mill owner that day with Jean when we visited the mill, and spent the better part of the summer trying to. Each day I called to try and arrange to introduce myself, he would be away, or at home, or with a customer. For the first month or two this was all right because we weren't in a rush to move, but as the summer started winding down it was starting to get rather important. I was getting the distinct impression I was being avoided. When I finally did contact him, it was quite clear he did not want the Optaphone in the mill. It's funny; I have always loved the candor of rural people. You can always sense honesty, and I was very disappointed that he hadn't just told me this straight out. I don't know what purpose stringing me along accomplished. Later, I discovered that at the time he was trying to sell the mill, and didn't want this equipment in there if it might have affected the deal. Just tell me that, I'm a big boy. I'll make other arrangements.

So now I was back to square one and had to find a location for the send unit. The mill was the best location in town but the more I thought of town, the less excited I was about it, since it sits at the bottom of a the Salmon River basin. All the land around it rises up, and then about 4 kilometers toward our house there is a large Canadian Shield escarpment with puts us that much higher than town. So from a line of sight point of view, town itself was not the greatest location. We could ask Faith and Mike Khouri, our nearest, and only full time neighbor on the escarpment, if we could put it in their house, but we felt that would be a huge imposition. Being from the city, we were still in that sort of "don't want to be dependent on my neighbor" sort of mind set.

I then began to research Jean's idea of a remote utility box to put the transmit unit in, and decided it looked like a viable option. But I needed a location. This is where I was willing to impose on Mike Khouri. He was as helpful as you could ever imagine. When I stopped into his place to ask, he immediately said, "Come with me" and off we went in his truck to scope out a location. Mike and Faith had moved here in the seventies, and had actually opened a very successful cross-country skiing destination. It had a chalet and many miles of ski trails that he hacked out and kept groomed. So Mike knew his hundreds of acres extremely well. After showing me where to be careful because of trails the deer used and often crossed the road on, he pointed out a rise at the side of the road, which was the highest point around. The only challenge was that it wasn't on Mike's land, but on some land he had sold to a couple from Toronto.

So the next weekend that the owners of that land, Richard and Angela were up, we approached them about putting up a phone pole on their property, about 100 feet from the road. They were wonderfully gracious and said yes. Eventually they drew up an official looking agreement indicating obligations on both sides, to avoid any misunderstandings. We decided it would work well when the final obligation on our part read, "The Mathers agree to be nice to Richard and Angela." That sounded like an expectation we would be able to live up to.

So now, I was a person from the city, with no real skills, who had to get a telephone pole installed in solid rock, and figure out how to install a utility box with high tech equipment, powered by solar power. If I were ever outside of my comfort zone before, now I was doing a space walk without a tether strap.

Luckily I found a company in Deseronto that put up telephone poles. So one day in September after we had bought the place, I met the crew at the spot where the pole was going up. The hill was typical of our area, solid rock with some spots of marginal soil. The crew found a suitable place to put the pole, and decided they'd have to blast out the hole, with dynamite. Once it was ready to go they asked me to drive up the road, so that none of the rocks showered down and broke my windshield. I was ½ a mile down the road when I realized I had meant to tell them something. Today, I cannot for the life of me remember what it was, but apparently it was important enough that I drove back into the rock shower to tell them. As it was, when I got there, they'd already blasted out the hole and mocked me in their "just another citidiot" (an amalgamation of "city" and "idiot") way, which I was becoming quite used to.

They then maneuvered the pole into the hole with a backhoe and began back filling the rock they'd blasted out. As they were finishing up, I noticed a huge rock with a flat side on it. With the exceptional skill of a surgeon, they used the backhoe to set the rock against the pole, so it formed a perfect shelf for the phone equipment box to sit on. Well, it was almost perfect. A bag of ready mix cement and a few small flat rocks from the area allowed me to craft a perfectly horizontal base.

Then it was on to the communications company in Mississauga, which had quoted me the enclosure and related equipment. They had calculated the amount of electricity the Optaphone would draw, based on the worst-case scenario, which was two weeks of cloud in the winter. The box would be sitting outside, completely exposed, and cold temperatures reduce a battery's performance. The utility box would be insulated with blue rigid foam insulation, and the minor amount of heat produced by the Optaphone and batteries should keep the unit working all winter. The solar panel was 75 watts, and I remembered being surprised that the box said "BP Solar" when I picked it up. British Petroleum was then the largest producer of solar panels in the world, and had changed their name to "Beyond Petroleum." As their corporate materials said, "In 50 years we won't be in the petroleum business", so it was time to change the name. This was the first time I had an inkling that something might be up with the world's energy supplies.

Once I had all equipment ready to go, I asked the only person I knew with skills in all areas that were going to be required for this installation, my neighbor Ken. Ken owns spikes, which you strap to your feet and legs to allow you to climb a wooden telephone pole, but he hadn't done it in many years. But he was game so off we went. Ken scaled the tower, then I began passing up equipment via a rope. I don't know how he did it, but Ken was able to pull up a fairly heavy solar panel, and wrangle it into position to bolt to the frame on the pole. I will always remember Ken calling for a bolt to fasten something at the top. Once I found it and was ready to toss up, Ken asked how many I had. I said "One." I still remember his response, sort of like a lecture from your Dad when you're little. "So let me get this straight. I'm 35 feet up a telephone pole, trying to attach all this equipment, and you have to throw up a bolt, and if I drop it, it will fall into the rocks and underbrush and never be seen again? And you only have the one? Is this what you're telling me?" From that day forward, if I was working on a project that involved screws or fasteners and Ken was involved, even if it was in a garage with feet firmly planted on the ground, I always purchased a number of the items I needed. For the marginal cost of the extras, it was insurance that the job got done that day.

But Ken caught the bolt and the solar panel was attached, and the antenna pointed at our house. I was very excited about this new location, because it cut down the distance the radio signal had to travel, by almost half. If it had worked from the mill in town for Jean, it boded well for this location, especially since I was significantly higher than the town location. Now the challenge was getting the Bell line put in. Anyone who has worked with a large utility, be it phone or electricity, has a pretty good idea of how accommodating they can be… NOT! So imagine how easy it was to get a phone line put into a box, mounted on a pole, with no street number, with no structure around, where the bill was going to be sent to someone 6 kilometers down the road. It was just going to a phone pole. And yes, it was a challenge. Eventually I had to trick them, and arrange to meet the installer there to sort things out.

So we had the power to the box via the solar panel, the antenna pointed, and the Bell line in. All I had to do was test it. Now, I cannot imagine the excitement of Alexander Graham Bell when he made that first call, but after the hoops I had jumped through to get our Optaphone system working, I think I was pretty close. This was key to us moving to paradise. We couldn't move up until we had a phone, and now in October we finally had one. Eureka! We finally had a phone. One of the things

Cam with his laptop at the remote phone box, powered by a solar panel at the top of the phone pole. The box includes batteries and the Optaphone which converts the phone signal into radio frequencies.

that's so great about living off the grid, is that it can make such mundane things that one usually takes for granted, seem so important. This magic box that let us chat with the outside world, was my ticket out of suburbia.

I came up the following weekend with our fax machine, and it worked fine. And the weekend after that, I had set up an account with Kingston Online Services, and was able to get our Internet working. When I got home to Burlington that Sunday night the first words out of my mouth were "Let's put the house on the market and get outta here!"

As with any piece of technology, the Optaphone was not perfect. Since it used radio frequencies, it was susceptible to quality degradation during rainy weather, and if we had freezing rain and the antennas iced up, it wouldn't work at all. It also had an annoying habit of periodically simply packing it in during a phone conversation. The challenge with this was that to the person on the other end of the line, often a customer, it sounded like we were hanging up. It is a testament to the loyalty of our customers that they were patient with us and got used to us hanging up on them, and waited for us to call them back when the system had corrected itself.

The other major limitation with it was the Internet. It was very limited in bandwidth, or the speed with which it could connect. I think the fastest we ever got was 14.4 Kb, and it was usually 12.8 Kb, which is about a million times slower than the high speed many people are getting used to in the city. The Optaphone could only do one thing at a time, so if you were on the internet, you didn't have a phone. As our business became increasingly reliant on sending files, especially PDF proofs to customers, this became a huge challenge.

We had tried a number of things. I had rented a small office in town, where I could go and send and receive files. I could also surf the net, which was becoming increasingly important in our business, especially when we'd need a custom photo from an online photo service. The office was not high speed, though, so it was only 56 Kb and while it offered an improvement, it was only marginal and I still had to drive into town, which was not our goal in moving to where we had.

I knew it was time to look at a better system when I found myself sitting in a lawn chair, by the phone box with my laptop plugged into the Bell line directly, because I had a large number of files I had to send. Oh it was a lovely day, and the bugs were fine, but somehow, as unique an experience as it was, it wouldn't have offered the same appeal if it was January or had been raining.

So after a number of years of very limited internet access, we installed a high-speed satellite internet system. I often joked that I moved up to the woods to live the 1970 Harrowsmith Magazine/Hippy dream of getting back to nature and living simply, free of the bounds of civilization and technology. The solar panels and inverter that made my life here possible were very recent, extremely sophisticated pieces of technology. Within months it was obvious if I didn't get my daughters television, there was going to be a minor revolution, so a satellite went up for TV. And now, I was adding a second satellite dish to the house for high speed internet. It was looking less like my hippy commune dream and more like NASA Mission Control in Houston.

The satellite system was not cheap either. By the time we were done it was close to $3,000. This included the installation of the satellite dish as well as an up and down modem and a router. We use Macintosh computers, and the system was set up for Windows-based machines. The router didn't care what our computers were, and it allowed us to have multiple units connected using the system simultaneously. It was truly a godsend. It made our life infinitely better. First off, it allowed us to keep the phone free when we were using it. And its speed was like jumping from a bicycle to a Formula One racecar. Not only was everything easier, from sending and receiving files, we could now realistically use the Internet to do things we had been hesitant to in the past, like banking on-line. With the Optaphone being unreliable, if it cut out while you were conducting a transaction online, there was the concern that records would go awry. So this piece of technology changed our life, and helped us reduce how much driving we did, because we could now trust using the internet to do things online.

In the meantime, the Optaphone continued to be challenging. One time it stopped working completely. I took it to Tordiff Communication in Brockville to have it looked at. As was often the case with technical things, everything checked out there fine. I took it back home and put it back into the utility box on our phone pole in the woods. After much consternation as I stood staring at it in frustration, I noticed that the power cord coming from the batteries looked different. When I took it off and examined it closely, the power wire was corroded. Some moisture had built up in the box and dripped on to the cabling, and most of the threads had rusted away. Only a few were left, and they only worked when they felt like it, hence our intermittent service. Luckily I was able to get another cable fabricated and I was back in business. As I had learned from Ken, I actually got two fabricated, in case I ever needed it again.

Another time it stopped working, and again, it took me several trips before I realized the batteries were toast. Deep cycle lead acid batteries have a natural life, and after many years of being charged and discharged they had reached the end of theirs. And of course they were insanely heavy, just at my limit of carrying capacity. The challenge was compounded by the fact that I had to walk the olds ones down from a steep rocky hill, and then drag the new ones back up. Unlike the Himalayan Mountains, there is an abundant lack of Sherpas with exccptionally strong backs, skilled in the art of hauling heavy gear up steep inclines, in our area. Once I managed to get the new batteries up there and get the system working again, it was like I had reinvented the wheel again. There is nothing as gratifying as trouble shooting a problem and fixing it. It's why I think so many professions like being a repairperson are so underrated. Fixing something that is broken is an infinitely gratifying and soul pleasing experience. Pushing electrons around on a computer pales in comparison.

The final chapter of the Optaphone began one spring when we were having problems once again. I went to the phone box to investigate, and discovered that a colony of ants had decided to inhabit the box along with our high tech equipment. They had managed to find an entrance through one of the cable holes, where the caulking that kept water out had shrunk, and they had moved in, in a big way. I don't watch horror movies, but opening the utility box that housed my family's lifeline to the outside world, and seeing tens of thousand of ants living in there and running amuck was just about more than I could take. As usual, even though my tendency was to run screaming to the car waving my hands madly over my head, much like a 1950s horror movie, I tried to remain calm and stifle my impulse to throw up my hands in defeat. I had to think this one through. I had a problem, and I needed a solution. I have been eating organic food for 20 years and have shunned the use of any insecticides or pesticides in the house or on our lawns or gardens, but I drove right to town and came back with two of the biggest most toxic cans of "Raid Ant Killer" I could find and unloaded them on the unwelcome guests. Yes, I am a failure as an environmentalist. I expect Greenpeace will rip up and return my membership card when they read this.

Surprisingly, the ants didn't seem to have caused a problem with the equipment itself. They had burrowed into the blue rigid foam insulation, but the electronics didn't seem to have interested them. So it was another trip back to the repair shop, but this time the prognosis wasn't good. I took the unit back to our phone box and tried a number of things, including my patented patting myself on the head and turning around three times while chanting "Abra cadabra", but nothing helped. This is where friends in high places come in, or at least in well-positioned places. Our friend, Jerry, from Burlington, specialized in radio communications for a large phone company. He had heard of a system called a "Tellular" unit, which they were evaluating.

We researched the system and it looked very promising. While we were in a marginal cellular service area, there were towers not too far away. If we purchased a Tellular unit, we could add a power booster, which would amplify the unit's signal, and if we pointed the antenna at an existing cellular tower, we should get a good signal. The Tellular unit would then take the cell call, and basically turn it into a phone call like we were used to getting on a traditional phone, with a dial tone and the works. After our friend Bill Kemp read the specs and gave it his engineering stamp of approval, we ordered one.

It required a different "yagi" directional antenna than the one we had, so when I had the antenna company install it (yes, my fear of heights once again got the better of my pocketbook) I had them put on a rotor, like my parents used to have on their TV antenna in the prehistoric days before satellite dishes. This would allow me to rotate the antenna and find the best signal from any cell tower within a 360-degree radius. After the challenges we'd had with the Optaphone, anything I could do to improve the odds of success with the Tellular unit was important.

Once installed the Tellular unit worked very well. While the cost for minutes was more expensive, we no longer had the requirement of purchasing the license for our radio frequencies that the Optaphone required, and I no longer had to use the remote box, with its ongoing maintenance and costs for new batteries. The only problem we had was that it did not allow us to get messages. We followed the instructions provided by the cellular service, tried a myriad of ways, called their support line and after months of trying, finally gave up. Then about a year later we spoke to another couple living off the grid and they explained how they retrieved their messages with their Tellular unit. We tried it, and it worked fine.

This coincided with our growing belief that the world is getting too complex. Thomas Homer-Dixon has written an excellent book called "The Ingenuity Gap" with the thesis that the systems human beings are developing are getting so complex, we are having trouble managing them, and when things go wrong, they can go horribly wrong and cascade into a far greater problem than was conceivable in the first place. Technology is a double-edged sword for us. It allows us to live far from the crowds, and yet still be in touch and generate an income using computers. But the challenges the technology presents often detract from the goal of our move in the first place, which was to simplify our lives. The phone messaging system was a classic example of this. From the company that sold us the Tellular unit, to multiple levels of tech support at the phone company, no one could explain how to retrieve our messages. It was just a fluke that we eventually found out how to do so.

This played out in an even bigger way, just about the time the Opta-phone system packed it in. We had a large thunderstorm roll through the area. When it was over, several of our computers were damaged, as was our satellite internet system. As I ran around getting Ethernet ports replaced in computers, I also took our satellite receiver and transmitter modems to our dealer. He was able to get them working on his system, but had to change the setting so they would work with his satellite. The dishes with satellite internet have to be positioned very accurately for them to work properly. So at the dealer, it appeared the system was work-ing properly. But when I got it home and set it up, it worked, but very poorly. Sometimes it worked, sometimes it didn't and when it worked, it was extremely slow.

As luck would have it this was the same period when we were evalu-ating a replacement for our phone system. I had customers accustomed to communicating with me through email on the internet, but it wasn't working either. This was a phenomenally stressful time. I spent hours trying to troubleshoot the problem. The dealer did his best. I worked past the first level of tech support with the satellite company, and was even past tier two support to Level Three tech support, but no one could help diagnose this problem. Finally, one morning, a week into the problem, I woke up at 5 am. I suddenly realized that the dealer had changed the location on my laptop so that the satellite could locate my system when I was at his location. So I came out to the office, got into the setting software, found my original settings printout, keyed them in, and the system returned to working properly.

It was a truly terrifying experience to know that I was completely dependent on this system, and yet no one, at any level of support could suggest a solution. And when I realized that someone as technically chal-lenged as me had to solve the problem myself, that is when the terror truly set in. You're on your own in this world and don't assume some tech support person can help you. This is life off the grid.

During this time when we were without a phone system, we had a back up, which was a bag phone. This was a very old technology actually, from the days when cell phones didn't fit in your pocket, but you actually needed a bag to carry them around. These older systems also had more power working at 3 Watts, versus the ½ Watt of modern hand held digital phones. If you're in the city or an area with good cellular coverage, ½ Watt phones are fine. Where we are, the 3-Watt phone worked relatively well. We didn't want to use it exclusively though because it was such an old phone the cellular systems wouldn't support it, plus it used older analog service and many carriers were discontinuing this and going exclusively digital, which the newer phones used. The local phone company has since discontinued analog service in this area so that a bag phone doesn't work.

As luck would have it, during this period of time when our main phone system wasn't working and our internet system was barely functioning, our eldest daughter Nicole was at the far end of the province of Quebec on a French exchange program for five weeks. Nicole is an adventuresome daughter, and even though she had almost no French experience, she jumped fearlessly head first into the program anyway. But when she got there the magnitude of her challenge suddenly dawned on her, and she discovered the program did not allow English to be spoken, so it was sink or swim. So there she was, 10 hours from home, unable to speak French and stuck for 5 weeks with people who were not cutting her any slack. Normally she would call home for advice and reassurance but with all of our communication systems in disarray she ended up having to cope without her parents' help.

Of course I blamed myself for her predicament. When we began home schooling the girls when they were younger, French was my responsibility. I had taken French to grade thirteen and loved to regale people with my near "bilingual" capabilities. When we had started our business, we shared a photocopier with a French translation service. Whenever I used the photocopier I always engaged everyone in French conversations. It should have tipped me off that my French wasn't up to scratch when my visits became the highlight of their days and they'd gather round and laugh hysterically at my French. This was probably the first sign my French was marginal at best.

But I cannot blame the girls lack of French speaking on that. The first two weeks they were home, I sat down with them regularly and we worked our way through the French text we had purchased. By about the third week, when it came time for French class, there was a crisis in the business, which we operated out of the basement, so French class was skipped. Fourth week, another time sensitive job needed completing and by the fifth week my illustrious career as a French tutor was over.

So, if it wasn't stressful enough having no phone and no internet, when we were able to talk to Nicole on our bag phone, paying through the nose for each minute, my stress was compounded by my inadequacies as a French teacher.

Time was on our side though. Nicole got through the tough early days and ended up having a great time in Quebec. We got the Tellular unit installed and working, and we eventually got the internet back on line.

Anyone who looks at living off the grid with rose-colored glasses needs a lesson in reality. It can be a huge challenge.

Using our Tellular unit, every call is basically long distance. There is no wireless internet in our area and likely never will be. The high-speed internet service we use will probably never allow us to use VoIP. "Voice over Internet Protocol" is using the internet as a phone. When we were researching the updated "Renewable Energy Handbook" which we publish, we bought the equipment, but it will not work with satellite-based internet. You have what's called "latency" as the signal is bounced 26,000 miles into space and back, so talking to someone with this time delay is almost impossible.

So if you live in a developed area with phone service, make sure you spend a few minutes today being grateful for your phone. It's an amazing, inexpensive, piece of technology that adds greatly to your standard of living. If you're thinking about moving off the grid, get a handle on communication before you move. Try cell service first. This will be your best option. If that doesn't work, you can start researching point-to-point systems like the Optaphone. Or, you can be really hardcore, and rely on using the mail to keep in touch with the world. I know sometimes as the pace of technology accelerates, and my aluminum addled 50 year old brain tries desperately to keep up, I long for the day when I can let the computers and phone system die a slow death, and I can spend my days in the garden weeding my potatoes. Squishing an infestation of potatoes bugs is as big a challenge as I will want to deal with.

Cam and the girls at the phone box located in the woods at the end of the phone line. The solar panel at the top powers the Opta-phone unit.

6 Wife Off The Grid

by Michelle

It's one thing to have a cottage or weekend retreat that is off the grid; you expect to have more relaxed standards towards cleaning and cooking in one of those places. It's another thing to cook and clean in a home that is off the grid. How do you cook without an electric stove or a microwave? Can you run a vacuum? What about a dishwasher or a clothes dryer?

I know that no matter how committed a couple is to using renewable energy and living off the grid, most women have some doubts and misgivings about what it will really mean for them. (I know I am being sexist in suggesting that women bear the brunt of cleaning and housekeeping chores around the house, but many studies have suggested just that!) Whenever we offer Open Houses to folks interested in renewable energy, I always have one woman who takes me aside and says something like "Okay, what's it really like to live off the grid?"

The first thing that I did as I prepared to move into my solar-powered home was to give away most of my small appliances. My friends and family gladly took my bread maker, toaster oven, waffle iron, microwave, hairdryer and various other gadgets off of my hands. Assuming that my electricity would be limited in my new home, I figured it was probably best to go back to more traditional ways of doing things, using my own energy rather than what came through a plug. (As we have improved our electrical system I have actually replaced most of these items as I have found that they do have a place in an off-grid home.)

The next thing we did was to ensure that every light bulb in our new home was a compact fluorescent, which use a fraction of the electricity of an incandescent light bulb while providing the same level of lighting. At that time (1998), compact fluorescent bulbs were not as commonplace as they are today and were actually quite a bit more expensive. Now, thankfully, they have become very easy to find and don't cost too much more than incandescent bulbs.

Our original fridge in this house was a propane fridge but a few years after moving in and after adding some new solar panels, we chose to switch to an electric model. Although we use propane in our cook stove, hot water heater and in a small space heater in our dining room, propane is not a sustainable fuel and so we work to continually find ways to cut back and limit our use of it.

In our previous home we used an electric stove. It was a hand-me-down from my parents and was almond in color, which was a major improvement over the avocado-colored stove that Cam's parents had handed down to us when we were first married. I had used electric stoves to cook on for most of my adult life, except for one period of time when an old home I was renting had a gas stove. I had really enjoyed cooking with gas but since I'd never been in a position to purchase a new stove, I'd never had the opportunity to choose gas again. After having our home tested for electro magnetic radiation and discovering that the worst source was our electric stove, I was really hesitant to do much cooking. (At least that's the excuse that I used!)

A typical electric stove has no place in an off-grid home. Using electricity to provide heat, whether for a stove, oven, hot water heater, or space heater, is just not an efficient use of electricity. And when your electricity is limited you wouldn't choose to devote it to one of these power hungry appliances. You'd also find that using one of these would drain your batteries pretty quickly! When we first looked at this house there was a large,

antique cook stove in the kitchen/dining room. It didn't come with the house however, so when we moved in we were faced with the decision as to what type of stove to buy. Not wanting to rush into the decision, we actually used our BBQ (propane powered) and a loaner stove for the first few months that we lived here. Believe it or not, you can actually bake a homemade pizza in the BBQ!

Eventually we decided that we needed to splurge and we bought a large, antique-looking cook stove to sit in the place of honour between the kitchen and dining room table. It is a spot that is visible from the moment you walk in the front door and so it just wouldn't have worked to put a little white box of a stove in that place. We found the perfect stove for us, a Heartland Oval stove, made in Kitchener, Ontario. We went really crazy and ordered a lovely forest green colored one with six burners. This particular stove is available as a wood-burning model, an electric or the gas (propane) one that we purchased. We did consider the wood burning model but my mother, who lived in an off-grid cabin with my father when he first returned home from the Second World War, advised us against the wood burning one. As she put it "you don't want to have to get a whole wood burning stove going just for a cup of tea. Especially in the summer!" Since we knew we were already going to have two woodstoves to contend with (one in the living room and one in our guesthouse) we decided to order the propane model. It is a beautiful stove and it receives many compliments from guests to our home. The only downside of owning such a large, six burner cook stove is that people expect you to use it and they expect you to be a good cook too!

This stove was the sort of luxury we rarely engaged in, even though it fit the house perfectly. Cam eventually came up with a rationalization. Rationalizations are important for big, irrational purchases. When we lived in the city and were both busy in our business and with the girls, Friday night was "Pizza Night." A delivered pizza would routinely cost about $20. When we did the calculations ($20 x 52 weeks = $1,040) we realized that we were spending over $1,000 a year on pizza. Now that we live so far out in the country, there is no restaurant willing to deliver pizza to us so Friday night became "Homemade Pizza Night." Since we were now saving $1,000 a year on pizza we figured that the stove would be paid for in less than 4 years. It's the stove that pizza bought!

Now our cooking has evolved even further along. During the winter months when our woodstove is heating our home, we make use of the top of the woodstove by always having a kettle (or two) boiling away.

We reheat soups and other meals on the woodstove as well. Recently I have begun using the inside of the woodstove too. Once the coals have burned down and the stove is just giving off a lovely warm glow, I wrap potatoes up in foil and place them inside, away from the hottest coals, to bake. They take about an hour, which is about how long they take in my cook stove, but no propane is used in their baking.

During the sunnier months Cam and I use a "solar oven," which is a glass-covered box, which uses the heat of the sun to cook or bake things. We had attempted to make one of these on our own, using ideas and plans from the internet, but in the end we discovered a brand from the U.S. called "Sun Oven" and this handy little box is just amazing! Made of advanced lightweight materials, lined with black colored metal, this unit has a glass lid. The box is also surrounded by a shiny metal collar that reflects light into the box. On a sunny day it takes very little time for the inside of the box to reach temperatures of 300 degrees Fahrenheit. We always keep a kettle inside the Sun Oven so we have almost instant boiled water whenever we want it. We reheat food in it and I've baked banana bread and granola in it. One summer we had a huge surplus of Roma tomatoes. We experimented with drying them in the Solar Oven and it worked!

Now that we have improved our electrical system so much by adding more panels, adding a wind turbine and improving the batteries and other bits and pieces of the system, we find we often have "excess" electricity on a sunny and/or windy day. We've gone back to using an electric kettle, an electric toaster, a toaster oven and induction stovetop on these days. Our aim is to use as little propane as possible for our cooking.

Cleaning is another major task in any household. Not only keeping yourself clean but your clothes, dishes, walls, floors and all of your possessions. Cleaning generally takes water – in some cases, lots of water. This can be challenging in any rural home, especially during a dry season, but in an off-grid home it is even more challenging as running the pump that pulls water up from a well can be a large use of electricity. When we first moved into our off-grid home we could always tell when our pump came on. The lights would dim during the initial surge and then return to normal as the pump resumed a more regular operating mode. In fact, as I mentioned in a previous chapter, if our batteries were highly charged and our pump came on, our inverter would actually shut off the power completely in order to protect our batteries from this huge surge in demand!

Thank goodness this doesn't happen anymore with our new improved inverter and charge controller, but that doesn't mean that we don't continue to be as water efficient as possible. After all, fresh water is in short supply in many parts of the planet and we should all recognize it for the valuable resource that it is.

One of the things that the previous owners of this house did to get around the problem of having a pump requiring a surge of power was to install not one but two pressurized water tanks. This means that as the pressure in the tanks decreases, indicating a low water level, our pump comes on and stays on long enough to fill two whole tanks. This cuts down on how often the pump turns on, thereby limiting the effects of the surge.

The water heater that was here when we moved in is propane powered. As we added solar panels and had more and more days in which we actually produced more electricity than we could use or store in our batteries, we chose to install an electric hot water tank as well. This is called our "diversion load" meaning that once our batteries have reached their full potential, we divert the excess power being produced by our solar panels into our electric hot water tank. This allows us to preheat our water with free, "extra" electricity before the water goes into our propane-powered tank. This is a win-win situation since it means we not only save on propane but are also able to use excess electricity to do so!

Recently we installed a solar thermal system, which uses the sun to preheat our hot water. This system is made up of a panel that we mounted on the roof of our back porch (facing south.) It required the installation of a third water tank and unfortunately the only place for a new water tank was in our bedroom! Yes we have a water tank in our bedroom. I call our design theme "Early Industrial" as not only is there a rather large, industrial-looking water tank taking up space but also the corresponding pipes and tubes and controller box too.

I was talking to Cam's grandmother one day on the phone. She is 99 years old, still sharp as a tack and living in her own home. I was describing Cam's latest exploits to her and the installation of a hot water tank in our bedroom. I was hoping for some sympathy from her but I laughed when her advice to me was "Oh, just put a curtain around it!"

Cam would like to devise some sort of piping system in which water travels through our woodstove for preheating, but luckily I have managed to discourage him from filling up our living room with copper pipes, at least so far!

We do use our woodstove to heat water though. As I mentioned we always have a kettle or two heating up on the top of the woodstove. I love hot baths and so if I am planning on having a soak I will fill large pots with water and heat them on the woodstove to supplement my bath water. Did I mention that I like my baths ***hot***?

Showers are more efficient than baths though and so I limit how often I treat myself to a soak in our claw foot tub. For water-efficient shower-ing we use a low-flow showerhead. We also use a dual-flush toilet and a low-flush toilet and have aerators on all of our faucets, which allow us to use less water and still get a useful stream of water.

Dishwashers can be used in off-grid homes but since I am convinced that my method of dishwashing is much more water-efficient, I have chosen to go without one. This isn't a big sacrifice on my part, since I've never owned one. I know that there are many studies that suggest that dishwashers don't use any more water than washing dishes by hand, but I often wonder who they are using for these studies – I use the lessons learned in Home Economics class many, many years ago and using a dish-pan in the sink allows me to wash a lot of dishes with a little bit of water.

We are careful about the cleaning products that we use. Instead of purchasing expensive, over-packaged and perfumed commercial products we use more natural ones such as vinegar, baking soda and borax. These natural cleaners are better for us and better for our septic system.

We do have a clothes washer. When we moved into this house we excitedly purchased one of the new front-loading washing machines, which use a fraction of the water and electricity of the traditional, top-loading machines. Not only do these new machines save you money, water, detergent and electricity, they are also easier on your clothing, allowing you to keep and wear your clothing for much longer. We had the new front-loading machine delivered to our house and waited expec-tantly as the technician hooked it all up. It didn't work. He tried all of the tricks in his repertoire but to no avail. He suggested that he would send another technician out the next day who was more familiar with these new machines. Technician #2 was also unsuccessful. Both of these men were stumped as to why a brand new washing machine would not work at our house. Finally a third technician was dispatched, and as we had with both previous technicians, we warned him that we were "off the grid". After his attempts failed as well, he asked what kind of inverter we had. When he discovered that it was not a true sine wave inverter, he announced that the sophisticated electronics of our new machine would not work with our existing inverter.

Since we weren't in a hurry to replace our inverter just then, we had our appliance dealer pick up the washing machine and bring the most efficient top-loading machine that they sold. We are still using that machine to this day, although now that we have upgraded our inverter and could now use a front-loader, we are tempted to replace our machine!

I do not have a clothes dryer, although there is no reason why I couldn't have a propane-powered one here. Instead I use a more sustainable way of drying our laundry - the solar powered way. During the spring, summer and fall, I use a clothesline (actually 3 of them) to hang my laundry out to dry. Not only does the wind and sun dry my laundry, but also when I bring them inside they have that "April Fresh" smell that you just can't get from a bottle of fabric softener. I try to dry towels on windy days when the breeze helps to fluff them up and make them soft. During the winter I use drying racks and often locate them around the woodstove. Again, this is a win-win situation, since while my laundry is drying it is putting much needed moisture into the air.

We use a clothes iron from time to time, but even before moving into an off-grid home I wasn't a big fan of ironing! In fact I will always remember the day I dragged out my iron and ironing board when my oldest daughter was about 3 years old and she asked quite innocently "What's that?" Apparently it had been some time since I had last used my iron!

As far as other electric appliances go, we do use a vacuum but more often than not I choose to sweep our floors and shake out our rugs and mats. This is when it pays to have easy-to-care-for hardwood floors and a minimum of carpeting. Needless to say, I generally wait until a sunny day to use our vacuum.

Performing all of the usual household tasks isn't that different in an off-grid house. As with other choices and decisions we've made we try to perform these tasks as efficiently as possible and purchase the most efficient appliances to help us. I like to joke that living in an off-grid house gives you the perfect excuse when unexpected guests drop in and find your house untidy – "Sorry but it hasn't been sunny enough to vacuum lately!"

7 Our Place In The Sun – Solar Power

By Cam

"On the spaceship superstar, I've got a solar powered laser beam guitar" **Prism**

Solar power is cool. It's insanely cool!
Using the sun to power your house and all your home's electrical needs is unbelievably gratifying. Using it for just part of your electricity is equally brilliant.

We had dreamed about using solar power for many years, but actually living with it as our main power source has given us a new respect for the high standard of living which the electricity that flows into our home provides.

There are three main ways that you can use the sun's energy in your home. The first is solar electricity. This is using photovoltaic or PV panels to power your appliances with electricity. The second is solar thermal. This is using the sun to heat water for your domestic hot water such as baths and laundry. The final is passive solar. This involves designing your house to use the sun's energy to heat your home. In some climates this will only meet part of your heating requirements, but in some southerly climates, it can just about do it all.

In our house, passive solar wasn't going to be much of an option, since the house had already been built. The amazing thing is that when they built it in 1888, they managed to position it so that the back of the house faces south. This means that on sunny days in the winter, the sun's rays coming in through the windows do help to warm up the house. It amazes me that in this day and age, I often seen new homes being built without any thought to taking advantage of the heat from the sun. People are more likely to build their homes facing a lake, or facing the road. Taking advantage of passive solar just seems like a no-brainer to me!

Basic passive solar design means having lots of windows on the south side of the house, so the sun will warm it up, and having few or no windows on the north side of the home. The north never sees the sun, and since much of the cold wind that blows in North America sweeps down from the north, the less heat loss you have through windows on the north side of the house, the better.

When people build ultra-efficient homes, like the type that they build into the side of a hill, they face the house to the south, while the buried part is on the north. Again, this way the sun will warm the living space in the winter, and you have less heat loss through the parts of the house that are covered in earth.

If you were thinking of adding solar panels to your home, the fastest payback would be for solar thermal, or using the sun to preheat your domestic hot water. Anyone who has turned on a garden hose that has been lying on the ground on a sunny day for 20 minutes is well aware of the sun's ability to heat water. I talk more about solar thermal in Chapter 13 when I discuss water.

Many of the visitors to our home ask if the solar panels keep our home warm. The solar panels they're looking at are photovoltaic panels, or PV. They are designed to convert sunlight (photo) into electricity. They are not designed for heat, and if you wanted to heat your house with electricity generated by your solar panels, you'd need almost a football field's worth of panels to do it. Even then, you'd still have the challenge of keeping your home warm on cloudy days. Also, the months when you want that heat, November and December, are often very cloudy and short.

One of the first projects we tackled once we got settled in at Sunflower Farm was to fabricate a tracker for our solar panels.

When we bought this house the solar panels were on a steel frame, on the ground, and permanently pointed south. The sun moves across the sky during the day and so with our panels in a fixed position, we were losing some of the potential energy whenever the panels were not directly facing the sun.

As I watched the sun track across the sky, I began to realize just how often the panels were not in a prime position to make the most of the sunlight. Pretty soon this began to drive me crazy and I looked for ways to correct the problem. We had friends up one weekend and after discussing the issue with them we eventually dragged a huge mirror out of the guesthouse and positioned it to reflect some of the sunlight that was now off the panels, onto the panels. It worked minimally, and even if it had worked well, it would have meant standing by the mirror and moving it all day. This did not sound like an optimal use of my prime income-earning years.

So it was time to ask our exceptional neighbor Ken for a hand. Ken had actually fabricated the steel frame that the panels were attached to now. So over the course of a few weeks Ken came up with a design for our tracker.

The first step was to source the required steel. While some commercial trackers use aluminum, ours was going to be made of steel. It was cheaper and stronger than aluminum, and Ken had the equipment to weld steel. Welding aluminum requires specialized equipment that Ken didn't have. Our first stop was at a steel supply company where the owners knew Ken well, and who thankfully had two perfect pieces of pipe for the base of the tracker. These would become the pipes that held the tracker.

The goal of a tracker is to be able to turn the PV panels to face the sun as it moves across the sky. So we started with two pieces of pipe. The largest one was going to be cemented into the ground. On top of it was a flange, which is a large, flat ring of steel with holes drilled in it. Think of it like a Frisbee sitting on top of the piece of steel pipe. If you have ever seen pipelines, you'll notice that you have sections of pipe that are attached together. They are either welded or bolted together with a flange. The two flanges at the end of each piece of pipe are bolted together, with the bolt passing through the holes drilled into the flange. The second piece of pipe, which was shorter and thinner, would partially fit down inside the larger pipe, and partially extend above it. The 3 feet that was inserted into the larger pipe allow the upper section to turn, while still being strong enough to support the tracker structure in high winds. To keep this smaller pipe in position, we had another flange welded on, at the 3-foot mark. This meant that the smaller piece fit inside the larger pipe, and at the 3-foot mark, the two flanges met. This kept the pipe in place and still allowed the pipe to turn.

To cement the pipe into the ground I had to dig a hole. The pipe, which was very large and made of steel, was 12 feet long. I wanted 4 feet of the pipe outside of the hole. That meant either cutting off part of the pipe, or digging a deep hole. In our part of the world, when you put something into the ground permanently, you have to get it below the frost line. This prevents jacking, which is when the freezing and thawing process of the frost forces items that aren't deep enough up and out of their position. Some people say 4 feet is enough, but I had decided it should be at least 5 feet in the ground to be safe. You also have wind issues, since the photovoltaic array is going to become a big sail, capturing the wind, and exerting extreme forces on the base. So, 12-foot pipe, 4 feet above and 5 feet below meant cutting 3 feet off. But that was going to be a big job. My philosophy in this case is always to avoid a big job by creating a bigger one. So I decided to bury 8 feet of the pipe.

Luckily the area where I was putting the tracker, near our battery room, is sand. Digging in sand is easy. Easy for the first 3 or 4 feet. By 5 feet it's more difficult, since the sand has a tendency to fall off the shovel as you're pulling it out of the ground. And if it's falling off at 5 feet, the vertical surface of the shovel is not going to allow you to drag the sand out from 5 to 8 feet. Plus most shovels have only 5-foot handles. So, my solution was a piece of a small tree that I cut from beside the pond, and, much to my wife's chagrin, a soup ladle. By 8 feet I wasn't that worried about making the diameter of the hole very big, just wide enough for the pipe. So there wasn't that much sand to remove. It was the same sort of action as when you see gondoliers dipping their poles into the waters around Venice, to push the boats forward, only less romantic. From this I learned, that you do not endear yourself to tough and rugged visitors, when your neighbor Ken insists on telling them that you dug the 8-foot hole for the tracker post with a soup ladle. You also don't come out looking like a rocket scientist.

Eventually I finished digging the hole and it was ready for the ce-

My neighbor Ken helping me mix the cement for the first tracker.

ment. You have three ways to get cement into a hole. The first is to get a truck to deliver it. They charge for delivery, especially out in the bush where I live, and have minimum orders, which I didn't require for this job. The second is to purchase "ready mix." These are bags that come premixed with the cement and aggregate. Concrete, consists of cement, often called Portland cement, as well as a course aggregate such as gravel limestone, and a fine aggregate such as sand. The combination is really important, so for a novice ready mix is an excellent option. Ready mix can be purchased at home building centers and is expensive, because they have done the mixing and have had to ship it. The bags are close to 90 pounds, and if you've ever lifted 90 pounds of dead weight, it can be challenging.

The final way, or Ken's way to fill a hole with concrete is to buy the Portland cement, borrow our neighbor Mike Khouri's cement mixer, and mix it yourself. So what I saved in buying ready mix, I made up for in driving to Mike's to borrow the cement, then getting a load or two of the sand we needed, and then dragging gravel off the driveway. Once you get the mixer going, it's like making bread dough. Ken would add one shovel full of cement, three shovels full of gravel, and three shovels full of sand. Then he'd mix in water from the hose and voila… concrete. We'd positioned the mixer beside the hole, so once the pipe was in place, we'd dump the concrete in to fill around it. This mixing process was repeated 4 or 5 times until the hole was filled. Then, we smoothed the top to make sure water will run off, and of course, we wrote our names and the date in the wet cement. This is a crucial part of a concrete job.

So, the pipe was in the ground and required a few days to set. We had the insert pipe, now we just needed to affix the solar panel frame to the insert pipe. Ken had fabricated a huge hinge, which would allow the panels to tilt at different angles in order to follow the sun as it got higher or lower in the sky, depending on the season. He had also added gussets, which are large pieces of steel designed to give the insert strength. We worked on all of this in Ken's garage. This is where I learned to weld. It's also where I learned that Ken's welds were the ones that people would see. My welds usually would involve welding scrap steel on the base of the tower to keep it from twisting. My welds ended up encased in 8 feet of concrete. But that's all right, because Ken is a much better welder and deserves the accolades for his work.

So now all that was left was to take the panels down to Ken's garage to weld the hinged insert section to the existing frame. We were able to get the panels on to Ken's trailer to take them the four kilometers down the road. Once there, Ken thought for a while about the best way to put it all together. After some thought, he grabbed a large washer, welded it to the top of the tracker, which is extremely heavy, then attached an overhead hoist to the washer and began lifting it into the air. Now someone familiar with the strength of a weld would probably have been pretty comfortable with this. But suddenly, I watched as my extremely heavy solar array, which included the panels that made the electricity for my computers, which I use to support my family, to make money, to put food on the table, hung, suspended in the air, over a concrete floor, held by a small steel washer. This is when I began to get my reputation with Ken for being a worrywart. Nothing fazes Ken. This fazed me, big time.

Ken was finished the welding late in the afternoon, and we reloaded the tracker back on the trailer. Then Ken gathered up his family members, son Brandon and son-in-law Jamie and we brought it back to our place. The PV array was heavy prior to the addition of the hinged tracker adapter. Now it was really heavy. Plus, we had to lift it 3 feet above the cemented base which stuck four feet out of the the ground. So it had to be lifted 7 feet in the air. This is where the back of a pickup truck comes in handy. We were able to lift it above the base, insert the inside pipe into the base, and drop it down inside. And it worked! It turned effortlessly with the help of the grease we had added to the pipes and flange. While it was no barn raising, we couldn't have performed the final assembly without the help of my crew of neighbors.

While preparing the base to be cemented into the ground, we also drilled holes and inserted grease nipples. These allow me to attach a grease gun and pump grease into the large base pipe, to ensure that the insert pipe turns smoothly and doesn't rust. Now, once a year a guy who makes his living publishing books on a computer, is out there with a grease gun, greasing up his solar tracker. Desktop publishing grease monkey boy!

Our tracker is not automatic like commercial trackers. Since I work in the office in the guesthouse, I am usually back and forth to the main house numerous times as the day progresses. Trips to the house for tea around 10 am, lunch at noon and more tea around 3 pm provide me with the opportunity to turn the tracker. I change the orientation toward the sun, and then put a steel peg with a handle into the flange to secure the two pipes so that the top one can't move in the wind. Trackers increase your performance more in the summer than the winter, because the path of the sun is so much wider than in the winter, but of course, you often have more power than you can use in the summer. If you were putting solar panels on your roof in the city, adding a couple of additional panels would offset the added electricity you'd generate with the help of a tracker, so there's no reason not to get solar panels on your roof! Working out of the house and having a neighbor as handy as Ken, allowed me to make a relatively inexpensive tracker.

We use our solar panels to generate electricity. When we purchased the house, the previous owners, Jean and Gary, had eight 60-Watt panels, for a total of 480-Watts worth of electricity. Jean and Gary did not have kids, so we quickly discovered the system was undersized. We also got a crash-course lesson in energy efficiency.

While converting to low wattage compact fluorescent light bulbs was easy, some components of our lifestyle weren't as easy to adapt quickly. Our computer use was one of the biggest challenges.

Most people who use computers are aware of the ramifications of changing computers, especially if their current model is working fairly well. "If it ain't broke, don't fix it" is the mantra we like to use. Using our computers for our electronic publishing business seemed to take this philosophy to the next level. Not only did we have the dozens of issues most users are aware of such as fonts, archived files, etc., but we also had a whole host of issues with fairly technical application software that sometimes worked together and sometime didn't. Once we had created artwork for a customer we would prepare the files so that they could be output to film ready for printing. Upgrading to a new computer made the job more complex. With the challenges we faced with moving off the grid, we weren't anxious to upgrade to a more efficient computer. It just meant too many other things to potentially go wrong.

Once again, that turned out to be a terrible mistake. We were using a massive color monitor for our work, with a large desktop computer. I have since given away the monitor and computer, but would make an estimate that together they used at least 300 Watts. If we were working normal hours, it would have been one thing, but being sensitive to our move 3 hours from our customers, I was more determined than ever to make sure we completed work quickly. So my days in the office were routinely 10 and 12 hours long. If your computer is on for 10 hours, and it uses 300 Watts, it is using 3,000 (10 hours x 300 Watts) Watt hours of electricity per day, or more commonly 3 kWh or kilowatt hours. On the grid, that's not so bad. Off the grid, especially with an undersized system, it's a bad thing. With about 500 watts worth of solar panels, if we got 6 hours of sunlight, that would produce 3 kilowatt hours worth of electricity. But that's on a good day. And that assumes you aren't using anything else in your house, like lights, or your water pump so water flows out of the tap when you turn it on.

So for the first year or so of our move off the grid, we had a huge deficit in how much power we produced, and we lived more "on generator" than "off the grid." This was quite discouraging, since our intention in living with renewable energy was to lower our impact on the planet. Our generator was a 10, 000 Watt ONAN, gas powered beast. It would not be classed as "efficient". It worked extremely well at charging the batteries, but it burned a lot of gas and therefore emitted a lot of greenhouse gases like CO_2.

Now I would argue that even running our generator more than we would have liked to, our ecological footprint using less electricity and creating some of it with solar power, was lower than if we lived on the electricity grid, and our electricity was coming from coal. In the United States 50 percent of electricity is generated from coal, and it is a major contributor of C02 and therefore greenhouse gases.

In the Province of Ontario where we live, a great deal of the power on our grid comes from CANDU nuclear reactors. The current government is looking to scale this up to 50 percent of our total power needs in the province. While nuclear reactors produce less greenhouse gases, their impact on the planet is highly suspect. They release radioactive tritium into the air and water. A study by Greenpeace recommended that pregnant women and children under 5 not live within 5 kilometres of one of the province's nuclear generating stations. Ontario's reactors produce 100 times more radioactive tritium than is allowed in Europe, and it is a known carcinogen.

Then you have the waste. Once the spent fuel rods are taken out of the reactor they sit in large pools on site for about 10 years to try and cool them down. Then they encase the radioactive plutonium in 60 cms of concrete and 2 centimetres of steel, and 10 years later they are still warm. And so we continue to accumulate these environmental time bombs with no permanent solution. There is lots of talk of an underground repository deep in the Canadian shield, but no location can be found because no one wants this potential hazard to life anywhere near them. Every 50 years these containers will weaken, and they will have to be torn down and rebuilt. This will continue on for at least 10,000 years, while the waste is still dangerous. What kind of legacy is this to be leaving to future generations?

In fact, one of the U.S. Department of Energy's challenges with their proposed Yucca Mountain nuclear waste repository is how to label the entrance to it once it is sealed up. We know that 10,000 years from now, it will still be a hazard, and we also know whom or whatever is on earth will not be speaking English. So we have to find some symbol, some universal representation of the danger that lurks within which shouldn't be disturbed.

So when I looked at my generator run time in this context, it didn't seem so bad. But I still wanted to reduce it significantly. So we started by looking at how many solar panels we had, and started to think about adding some, and at reducing one of our biggest loads.

So I took the plunge and started investigating laptop computers. Laptops use about 20% of the energy of large desktop computers, and with the size of a desktop publishing monitor, it might be even more. We were concerned about quality issues with laptop monitors for desktop publishing and photo adjusting. Luckily most of our work did not require extremely high end photo retouching, and the new generation of Apple laptops early in the new millennium was more than adequate for our requirements.

Getting our first laptop home and using it for a few days was one of those "eureka" moments for us. We had one of those "why didn't we do this a year ago?" moments. Oh yea, we didn't want the hassles. But we got it working and it made an absolutely huge difference. We cannot emphasize enough the importance of energy efficiency for someone living off the grid, and just as importantly, for someone on the grid trying to reduce their footprint on the planet.

The laptop helped us to reduce our loads dramatically, but it was still apparent we had a demand that was higher than the supply. Having teenaged daughters didn't help. Yes there was their computer and TV time, but there were also the showers. Never having been a fan of long showers, it was hard for me to figure out how a person could spend 45 minutes in a shower. I guess if I paid closer attention to the 20 bottles of products that had to be applied at different times, in different areas to accomplish specific body washing tasks, it would have helped. But the water for those showers had to come from somewhere, and to pull it from our dug well required a pump that loved to suck up power. If you were in the battery room when the pump came on, it sounded like the inverter was going to jump off the wall it used so much juice.

So we made the decision that it was time to add some more panels to help with this supply deficit.

This is when we discovered the wisdom of Ken building such a strong steel system for our tracker. It was solid enough to add on to. So within a year or so of building the tracker, it was time to get Ken thinking about an addition. Then we waited while the request gestated with Ken. In no time he came up with a system of using angle iron steel, which he would weld up at his place, then bolt onto the existing tracker.

Ken has this incredible ability to visualize what a final product will look like. He took measurements, and bought the steel. Then we got down to the tasks of measuring, cutting the steel, and welding it together. To cut steel, we generally used a grinder. When you cut steel with a grinder, when the grinder disc hits the metal, it sends a glorious spray of orange sparks cascading great distances. It's the sort of image Hollywood uses to indicate an industrial setting and it's one of those cool things guys like to do in a garage. Not having been raised in a culture that provided me with a specific rite of passage to manhood, I considered using a grinder to cut steel for our tracker to be more than sufficient. I could use a power tool to make a wave of red-hot sparks combined with a deafening roar. I was a man.

Once the steel was cut and welded, Ken hauled it up with his winch and we painted it with green rust paint. Then it was just a matter of hauling it to our place on Ken's trailer and bolting it to the existing tracker. This all sounds easy enough, but the addition was big and heavy, and had

Roy, Michelle and Cam's Dad hold the tracker addition while Cam and Ken bolt it to the back.

to be lifted into place. I had convinced my father to help and Ken had his friend "Cousin Roy" ready as well. Cousin Roy is not really Ken's cousin, but Roy says that since they both have a mother, that makes them cousins. So Ken arrived and shared an idea he'd had which involved changing the existing tracker so that it moved more easily. To change the position of the panels relative to the horizon, we had to unbolt a steel rod and it was very hard to move. Ken came up with the idea to cut a section of the steel out and replace it with a car jack. I, of course, just wanted the extra tracker section on, but as always Ken insisted that I learn some patience while he worked his magic. And now years later, as I effortlessly use the car jack to change the panels position every couple of weeks, I am grateful for Ken's wisdom. I just wish it wouldn't always come when I wanted to get something else done.

Once we bolted the tracker addition on, we were able to put our 4 new 75-watt Siemens panels on. These were wired up and we noticed a difference with the additional 300 Watts worth of electricity right away. When the sun was out, the batteries were charging faster. With the loads reduced by switching to a laptop, and the additional panels, life was getting easier, and we were running the generator much less often.

We had looked down the road a little bit, and had designed the tracker addition to have room not only for the additional four 75 Watt panels, but for 8 more panels. We didn't have the money to buy them at the time, but we had a feeling we'd want to add more in the future. As it was, the future was not that far off.

Many off-gridders switch fuel sources to make sure that their major heat requirements don't require electricity. In rural areas, they often use propane for this heat. Propane is like natural gas, but is in a liquid form, so it can be delivered in large trucks and stored on site in large tanks. Propane basically boils when it hits air, so as your appliance calls to the propane tank for fuel, some liquid propane boils and becomes a gas, which your appliance can burn, similar to natural gas. Propane is fairly close to natural gas, but you will still have to purchase an appliance designed for propane, or have a natural gas model modified to burn it.

In our house, our major sources of heat, which included our cook stove and our hot water tank, were propane. As we discuss later, we heat our house with wood. In the early days of off-grid living, when people accustomed to city amenities moved to the woods, they still had a desire for luxuries like a refrigerator. Some would argue that a fridge is a necessity, but I use the point of reference that humans have lived without refrigerators for centuries. So when people began trying to run a fridge off the grid, they discovered that they didn't realistically have the technology in terms of solar panels and wind turbines, to generate the electricity needed to run it. So instead they installed a propane fridge. This is a fridge that burns propane to start a chemical reaction which cools the inside of the fridge.

Propane fridges are very common in rural areas where there aren't always power lines, like remote cottages and hunting and fishing camps. Propane fridges can also be very dangerous. They are a source of combustion and they create carbon monoxide. Many of us now have carbon monoxide detectors in our homes, in case a stove or furnace doesn't work properly and starts venting this lethal gas into our homes. People still routinely die of carbon monoxide poisoning from using devices like propane fridges. Part of the lure of the remote experience is leaving so many of the restrictions of urban life behind. Some of these are good ... like cell phones and television. Some are not so good, like carbon monoxide detectors. So having a propane fridge venting into your home has its risks.

When we had a propane fridge we were completely unaware of the danger it posed. All we knew was that periodically it produced a horrible smell, and we'd have to turn it off, take apart the flue pipe at the back,

The first tracker with room for four more panels.

and wash it, and put a long pipe brush down it to clean out the soot. It seemed to us, that even though there weren't clouds of black smoke wafting from the fridge, that it probably wasn't making a positive contribution to our indoor air quality.

So we decided it was time to get rid of the propane fridge. But this wasn't without its ramifications. An electric fridge was going to greatly add to our electricity demand, and with our recently added panels improving our quality of life, we weren't anxious to now add another large load to the equation.

Luckily we had built space for 4 more panels on the tracker addition, so we could accommodate the new ones conveniently. Most people when they buy a new fridge simply look at the price in the appliance store, $500 or $1,000, and that is what they need to spend. In our case, we got to add on the price of buying more solar panels. We certainly didn't begrudge it, because it's the right thing to do for the planet. We were displacing the non-renewable, greenhouse-gas-producing propane, with clean solar power. But we were going to have to budget for it.

The first 75-Watt panels that we purchased were approximately $700 each, so we were paying about $10/Watt. So with tax, 4 of them were close to $3,000. For many people, with electricity rates, it would take a long time to justify a purchase like that. For us, it meant a much easier time in terms of energy production, and less generator run time, so it was easy to justify. For the panels we needed this time, we got together with several other off-gridders we knew, including someone who was putting in a brand new system. When we pooled our solar panel order together it was close to $35,000 worth of equipment, so we were able to get these 75-Watt panels for $560. It was great to see the price dropping.

Now we needed a new fridge. As fate would have it, our friends Bill and Lorraine Kemp were in the process of upgrading their fridge and so their small electric fridge was for sale. When they had moved off the grid 10 years earlier, they had researched and bought the most efficient fridge they could find locally, which was a small Sears fridge that used approximately 400 kWh (kilowatt hours) per year. It was ideal for us because it fit in the same amount of space as our propane fridge, which was recessed in the wall beside the pantry. There are some extremely efficient refrigerators you can buy from a company called Sun Frost in California. One of the reasons they use so little electricity is because they use massive amounts of insulation on the outsides to keep the cool air in and the hot air out. The downside of that fridge for us was that it would be 50% wider than the space we had. Also, it was almost double the cost of an average efficient appliance store fridge by the time you got it to Canada. It also meant that if there were problems it might have been more difficult to find parts and expertise locally to fix it.

So we bought Bill and Lorraine's used fridge. They had discovered that in the 10 years since they had bought the smaller fridge, they could purchase a regular-sized fridge, with automatic defrost, that used the same amount of energy. The automatic defrost becomes an issue in an off-grid home. Anyone used to a non-frost free fridge at the cottage, knows that after 3 or 4 weeks of use, the freezer very quickly becomes an "ice box", and it becomes difficult to fit things into it, as ice and snow buildup has displaced the room formerly used to keep things frozen. A frost-free refrigerator will go through an automatic defrost cycle regularly, where it basically uses an electric heating element to melt the ice or frost buildup. This can be quite a power draw, but modern fridges are becoming so efficient this is less of an issue. We have states like California to thank for these innovations. They have made it mandatory for manufacturers to steadily improve the efficiency of their appliances to reduce the electricity they require, and hence to reduce the impact on California's air, of burning fossil fuels to produce electricity.

As with so many seemingly simple tasks, like buying a fridge, we always have to go into research mode, to find out what an appliance draws

The first tracker now filled up.

and determine how it will impact on how much electricity we produce. It's the ying and yang of living off the grid, the supply and demand equation, the concept of resources being finite. For someone who believes that the planet is finite, it's the perfect way to live.

The amazing thing about living off the grid is that it teaches you about the word "finite". In the city, or on the electricity grid, power is infinite. You can use as much as you want, and apart from the odd disruption, it just keeps on coming. Without end. Living off the grid and generating your power with solar and wind, you discover that electricity is indeed finite. You have a certain allotment each day that you hope to live within. Some sunny and windy days, you have way more than you can use, and those are the days when you wash your clothes, vacuum, boil your water in the electric kettle, and live pretty much like someone on the grid. Some cloudy days, you let the dirty clothes pile up, heat the kettle on the propane stove, sweep the hardwood floors, and try and convince everyone to sit down and read after dinner, rather than firing up the television. If you have a good battery bank, you shouldn't really have to worry about changing your lifestyle for a couple of days, and then, many who live off the grid will just fire up the generator to run loads and charge batteries.

We have always tried to have our behavior mimic the weather, and undertake our most energy intensive activities on sunny days. Humans have lived this way for eons, so we don't feel it's a great inconvenience. Farmers would have to wait until the fields were dry enough to plant. There was no use washing clothes on a rainy day, because they wouldn't dry on the line. And it was best to harvest crops during a dry spell, to keep them from getting mould and spoiling prematurely. It is only in recent history than humans have had the luxury of limitless energy to smooth over the ups and downs of weather, and live their lives completely removed from any influence by nature. With an indoor washing machine and dryer, you can wash and dry your clothes in a very short amount of time regardless of the weather.

Lots of people who live off the grid have dryers. They use propane as the non-renewable, greenhouse gas belching fuel source. But we don't have a dryer. When we lived in the city we had a dryer, but we rarely used it. The clothesline did a fine job. And here in the country, the air seems even cleaner, even though so much of North America has poor quality air, from coal fired electricity generators and various other sources. As a man, I'm not proud to admit that Michelle claims that I am "clothes hanging challenged", and therefore the clothes hanging responsibility falls on Michelle. During the winter, once the air gets too cold to get the clothes to dry outside, we dry them on drying racks inside. Wood heat can seem dry sometimes, so having the clothes release their moisture into the air is always welcome. In 3 or 4 hours, the clothes are dry and the air is humidified.

With the new electric fridge installed and new panels added, we were getting much closer to the sustainable lifestyle we had been imagining when we moved off the grid. With the initial eight 60 Watt panels (8 x 60 = 480 Watts) and the two sets of four 75 Watt panels (2 x 4 x 75 = 600), we now had just over 1,000 Watts of photovoltaic panels generating our electricity. This was a eureka moment for me. As someone who had been challenged by all the electrical terms I had to learn on our solar odyssey, it was often difficult for me to conceptualize just how much power we had. Now, with 1,000 Watts or 1 kilowatt of PV, I knew if we received 1 hour of sunshine on the panels, we would produce 1 Kilowatt-hour of electricity. There are losses and efficiency issues involved, such as how much of this energy your batteries can actually absorb, but it was much closer to being something that I could get my head around.

We used about 5 kWh (kilowatt hours) of electricity per day. If the sun shone for 5 hours we would have enough energy to power our home for the day. If it shone for more than 5 hours in the summer, we had more power that we could use. If it shone for less than 5, we had to rely on the energy stored in our batteries to make up the deficit. And if the sun stayed behind the clouds for 2 or 3 days, we could either go to bed when the sun went down and bathe in the pond, or fire up the fossil fueled powered generator.

The average family in our province of Ontario uses 35 kilowatt hours of electricity per day. So we are very proud of having our loads so low. We accomplish this through energy efficiency. If you live on the grid and receive your power from a large centralized generator plant, you can also use energy efficiency to reduce your carbon footprint and your impact on the planet.

After several years of having 1,000 Watts of photovoltaic power, we decided we were ready to take the PV plunge again. The information available on climate change was getting more dramatic, and the Intergovernmental Panel on Climate Change was making it very clear that the climate was changing and humans were responsible for it. During those times of the year when we did not have enough sunshine, we had to make up the difference with our fossil fuel-powered generator. It used gasoline and it released carbon dioxide and we wanted to cut back on how often we had to run it. By adding more solar panels, we'd have to run our generator less often, even during the months when the days are shorter. This time we purchased much larger panels that were 165W each. They also were 24V rather than 12 V as our previous panels had been. These panels were more than twice the physical size of our 75W panels, and twice the weight. Once again it was Ken to the rescue, this time with an even more enhanced design. Now that he had built one system and seen how it performed, he could design the new tracker from the ground up to be even better. The bonus was that Ken had a friend Robin who also needed a tracker for his solar panels, so this time Ken would be fabricating two trackers. This meant economies of scale for both. Ken and Robin set about fabricating these when we were right in the middle of a very busy time at Aztext Press, getting our book "Biodiesel Basics and Beyond" to the printers. So they basically built me a tracker. The great news is that Robin and Ken have turned this into a business and are now fabricating trackers commercially. There is nothing more gratifying for people as passionate about renewable energy as Michelle and I are, to see the

The first four 165 Watt panels on the second tracker.

business grow as so many other people start integrating it into their lives.

The new tracker was being made in the fall, and wasn't ready to go until December. We had been incredibly busy and frankly, I was relieved that I wasn't going to have to install it until the spring, because I wasn't going to be able to cement the post into the ground until the ground thawed. It was frozen solid in December, as it had been every year on record in our part of the world. Then something incredible happened. Gradually, little by little, the winter started to warm. We were able to skate on Christmas and Boxing Day, but then we got a thaw. The ice turned to slush. But rather than a short thaw lasting a couple of days, this lasted weeks. In January, typically frost bitten Canadians were back to wearing their fall wind breakers. And while many rejoiced in the warmth, the overwhelming attitude of people was that it was "scary." This wasn't a short winter thaw. This was different. This was unprecedented. And as welcome as the warmth was, at its most basic level, Canadians knew something was amiss. Suddenly this theoretical, hypothetical, hard to conceptualize concept of global warming was starting to hit home. It wasn't good for people reliant on cold temperatures, like ski resorts that were laying off staff in record numbers as their business evaporated, but it also wasn't good for nature. Trees and plants that should be buried under a blanket of snow, and whose cellular functions should be slowed to a hibernation mode, were confused.

So I found it quite ironic, that the solar tracker that I wanted to install to reduce my CO_2 footprint, I could now cement into the ground, because the ground had thawed in January. I could actually dig a 6 foot deep hole in a month when traditionally I couldn't have put a pick axe through, because it would be frozen as solid as stone. And so, once again Mike Khouri's cement mixer followed the well-worn trail to my house, and I cemented another solar tracker pipe into the ground. After it set for about a week we had a crew over to drop the new tracker onto the pipe. Then Michelle helped as I bolted the new panels on. We had predrilled the holes for the panels while it was on the ground, which made it infinitely easier to attach the panels. It was just a matter of lining up the holes on the panels with the holes on the tracker steel, putting a bolt through, and then tightening a nylock nut onto it. A nylock nut has a plastic gasket inside which makes a very firm fit, and will not loosen under vibration that might occur as the tracker is shaken in high winds.

So now, with the addition of four 165-Watt panels (4 x 165 W = 660 W) to our previous 1,080 Watt tracker, we had over 1,700 Watts worth of solar panels. This was another red-letter day and with the addition of our new charge controller we were better able to track how much power was coming in and how much energy we were storing in our batteries. Ironically, solar panels don't like to be hot; they are more efficient in the cold. This is why if you install them on your roof, you should make sure they are on a racking system, which allows air to circulate under them. If they were affixed tightly to your roof, they would not produce as much power on a sunny day. With our trackers as vertical as they can be in the winter, and a blanket of snow in front of them, the panels are the happiest.

There is always a discrepancy with the rated wattage of the solar panels and what you get, but now on a brilliantly sunny day in the winter, when it's cold, the air is clear and clean, and there is snow on the ground, my solar charger tells me that we are producing over 1500 watts worth of electricity. Even on a December day with short days, we have produced more than 6-kilowatt hours of electricity, enough to power our home and bank some for the next cloudy day.

As always, when you build a tracker and leave room for more panels, it's hard to resist putting them up. Within a year we installed another four 175 Watt panels. By then Sharp had changed the design again, so my solar trackers always look like a hodge podge. But I see it more as a process of "continuous quality improvement." We just keep making the system better. Now with almost 2,300 Watts of PV we sometimes make about 17 kWh of electricity. And with all this extra electricity we've been able to cut down on our propane use significantly. Now our counter top has an electric kettle, 4 slice electric toaster, and a convection toaster oven, a microwave, an electric waffle iron, and an induction cooker. And now for much of the year we cook with electricity.

After years of fine-tuning, and upgrading, and adding panels, and building trackers and adding even more panels, we feel we are finally in the "groove" of solar power. It took us almost 10 years. It took reading, and learning, and experimenting, and spending money. But we are very pleased to have finally arrived at a place where the system is functioning extremely well. The fall we added our new wind turbine to the upgraded photovoltaic system we only had to run the generator twice, versus 15 to 20 times in previous falls. With today's gas prices, it would have been far more cost effective to simply run the gas-powered generator to make up the deficiency and create the greenhouse gases that go along with it. But we didn't. We spent the money and did the right thing for the planet.

And you need to do the same. You need to reduce your use of electricity. You need to replace lights bulbs and buy efficient appliances and use all of the techniques that Bill Kemp has shared in "The Renewable Energy Handbook".

Then you need to buy some solar panels. You need to start making your own power, leaving a legacy to your children and grand children. The solar panels will come with a 25 to 30 year warranty. With current electricity prices the panels will be paid for in about 15 years, and then after that, you have free power for life. If electricity is more than the 12 cents/kilowatt that Ontario power consumers pay, your payback may be even faster. If there are government incentives for the installation of renewable energy equipment, the payback could be faster still. But it's time you did it. It's time you stopped just talking about saving the planet and start investing some real money in it.

Our two trackers as of 2011, showing how upgrading our system has been a continous process.

8 The Kids Are All Right

by Michelle

Twenty-five years ago when Cam and I made the decision to have a family, we hadn't given much thought to the overpopulation of the world and the peril that our planet is in. I think we naively thought that everything would turn out okay. Today, I'm not so sure and I might make a different decision about bringing new children into the world.

Having had children, Cam and I felt very strongly about raising our daughters in a sustainable fashion. One of the first "enviro" decisions that we made was to use cloth diapers as much as possible and limit our use of disposable ones.

Cam and I have often remarked to each other that we can't seem to refrain from questioning the status quo and always finding a different way of doing things. Living in an off-grid house and eating a vegetarian diet are not "conventional" and the decisions that we made in terms of raising our daughters certainly followed in the "alternative" path that we had been on.

Even as we prepared for the birth of our first child, we were questioning the conventional methods of childbirth in which drugs are provided to the mother for pain. Along with our prenatal classes that we took with the hopes of educating ourselves as to what was to come, we researched ways in which the pain of childbirth was handled in other countries and we came across a T.E.N.S. unit. This is a small electrical box, a Transcutaneous Electrical Nerve Stimulator, which provides a small electrical current to specific muscles, which helps your body deal with pain. After getting the consent of both my doctor and the hospital's staff, I used the T.E.N.S. unit for both of my labours.

Once the girls were born we were faced with decisions around vaccinations. At the time it was common practice to begin the immunization program when the infant was just 2 months old. Cam and I spent many hours at the local medical university reading up on studies concerning immunizations and we made some decisions that were definitely out of step with conventional practices. First off we postponed all vaccinations until our daughters were much older and we felt that their bodies would have a better chance of dealing with them. We were also very choosy about which vaccinations they were given and, in fact, completely avoided the pertussis vaccine (for whooping cough) since it had been implicated in some adverse reactions. We felt that since our babies would be raised in their own home and their contact with other potentially sick people would be quite limited, their chances of contracting whooping cough (or any of the other illnesses) would be small. To this day, I am still quite ambivalent about the vaccination decision.

After spending two years at a neighborhood cooperatively run preschool our girls attended a local public school. At first we used our bike trailer to take them back and forth between home and school and as they grew older they were able to ride their own bikes alongside us. From time to time they grumbled about having to ride to school when it seemed as if every single one of their friends was given a ride in a car (or more often a van or SUV) but we saved trips in the car for days with the worst weather.

Just as I had been very involved in the girls' preschool, I was very ac-

The girls skating on a frozen pond behind the house.

tive in their elementary school, running a "Green Club" during the lunch hour and also as an active member of the Parent-Teacher committee. I had been an elementary school teacher myself for a few years before the girls were born so I felt very comfortable in a school environment. Or at least I should have. Instead I was noticing a "vibe" at the school that I just didn't like. We were also noticing a change in our girls with each passing year that they spent at school. They had entered their Kindergarten classes with a "spark" and a zest for learning. As each year went by though we noticed that their spark grew dimmer and they showed much less enthusiasm.

We had read about "homeschooling" and had suggested to each other that once we were living "out in the country" maybe we would give homeschooling a try. Finally a few incidents inspired us to move up our plans and offer our girls the option of being homeschooled while still living in the city. Nicole was eight years of age and in Grade Three when she reported that she was being teased because she wasn't shaving her legs yet. What? I was shocked and after questioning her further it became pretty clear that the other girls in her class were in a big rush to "grow up" and were spending their recesses talking about boys and make-up, etc. Nicole had stopped wanting to wear any clothing with cartoon characters on it, for fear that she would be teased. Katie was in Grade One at the time and while she was happy socially, she wasn't being challenged academically. My girls had always been big fans of "word searches" and I often made up word searches for them during long car rides, etc. One day Katie came home from school and announced that she was "sick of word searches." I asked why and she said that she had been given a word search to do six times that day – every time she finished her work earlier than her classmates she was told to sit quietly at her desk and do a word search. The only problem was that she was given the same word search… six times! No wonder she was bored!

At that point I decided to offer both of the girls the opportunity to be educated at home. Cam and I were running our electronic publishing business out of the house at that point so it wouldn't be a problem having the girls at home and spending some time each day with them on schoolwork. Both of the girls jumped at the chance. We assured them that each year in August we would let them reconsider their decision and they would be free to return to school if they wanted to. For six years they both decided to be educated at home. It wasn't until we had moved to our country home and they both reached those teenage years when friends become more important than your parents that they both decided it was time to head to high school. They put up with a horrendously long bus ride there and back each day. It is a testament to how strong the desire to socialize is at that age that they were willing to put up with all of the inconveniences of attending a high school that is 40+ kilometers (25 miles) away!

Many people have asked us about our decision to move from the city to the country and how it impacted our children. I've had people tell me that they are waiting for their kids to grow up and move out of the house so that they can finally move to their dream location. I always ask "Why are you waiting?" and inevitably I am told that their children would "hate" to move and they just couldn't put their children though something like that!

Our daughters were 10 and 12 years of age when we undertook our move from the city to the country. While they weren't thrilled with the prospect of living 3 hours away from their friends and our extended family members, they seemed to look at our move as a grand adventure. They knew that they would be able to have more pets out in the country, perhaps even some "big" pets. They knew that the house was bigger than our current one and so they would be able to have separate bedrooms. There were a lot of reasons for them to be excited by the move and Cam and I never once considered waiting until they were out of the house before we pursued our country living dream.

When we lived in the city, Cam often rode his bike to the top of the Niagara Escarpment where he could see from Toronto to Niagara Falls. On a Monday the air would be fairly clean but by Friday he could see the noticeable dividing line between blue sky and the brown haze that passed for air. This was what our daughters were breathing everyday!

We were anxious to get them out of the "dirty" city air and provide them with a place where they could run relatively freely. Luckily they were old enough to know better than to get lost in our woods or to fall into one of our ponds and in many respects I believe that they were at a prime age for a move like this.

Our daughters always seemed very flexible about the various changes and issues that came with our alternative lifestyle. At the new house Nicole had a separate bedroom with a door. Katie had more of an "open concept" bedroom. Jean and Gary (the previous owners) had removed most of the walls upstairs to turn the area above the living room in to a large den. We put bookshelves across half of the room to make a partition for Katie's bedroom. It wasn't overly private, or quiet, but Katie never complained. Eventually Cam built a permanent wall, with a real door that closed, so Katie finally had her own bedroom. I think that sometimes this type of adversity for children helps build flexibility for upheaval later in life.

There is no question that there have been many times when our daughters lamented the fact that we were "different" from the rest of society. My daughters don't eat meat, they were educated at home for six years and then they moved to an off-grid house. It wasn't until our older daughter was at a party while she was attending university and someone said to her "I hear you grew up in an off-grid house. That is so cool!" that she came to realize that other people actually found our alternative lifestyle kind of neat. In reality both of my daughters have told me that they look back at our homeschooling years with a great deal of fondness and they didn't appreciate the freedom until they no longer had it. Such is life!

Funny enough both of our daughters (now in their early twenties) are

Katie getting a riding lesson from our neighbor Alyce.

living in Canada's largest city and show no signs of considering a move back to the country. They love life in the city, take public transit to get around and they seem to take advantage of all that the city has to offer. I am happy to be able to provide them with a country retreat whenever they need some down time!

9 Blowin' in the Wind

by Cam

"Now the sound of wind chimes means power being made by the wind turbine!"

Cam Mather

We have noticed that there is a distinct male/female or masculine/feminine divide with renewable energy. We first noticed it while participating in renewable energy fairs where we sell our books. We often have a small wind turbine set up on display, as well as a solar panel.

Men will come right to the wind turbine. They are drawn to it like a moth to a light. They will spin the blades. They will spin the blades and thrust their arms into the path of the blades to see just how much it hurts. They try and figure out how it works. They'll ask how high it should be. They'll suggest they have an old alternator in the garage and they're going to make one themselves. Men like wind turbines because they break, and they will have to get out their tools to fix them. Wind towers are dangerous, whether you're climbing one or using a tilt up tower, there's always an element of danger there. Men like danger. You're much more likely to hurt yourself putting up a wind turbine, so men like wind turbines. They are loud, and tall, and noisy and show off. A wind turbine screams out "Look at me, I'm here, I'm important." Wind turbines make a statement.

Women on the other hand, are drawn to the solar panels. Solar panels are quiet and unassuming. They work away, producing clean electricity, patiently, waiting for the sun to warm them. They do not scream for attention. They are happy to take a back seat to a wind turbine, even though they are often much more productive. They don't break down, they just keep working away. They just get the job done, and they don't need a lot of fanfare to do it. No big fuss. Just let some sun shine on them, and they'll make electricity, but you won't even know that they are there.

As a male feminist, I've always enjoyed my solar panels. I think they're awesome and I love what they do. But because of some deep-seated, Cro-Magnon DNA impulse, I am drawn to wind turbines. And after erecting a brand new tower and wind turbine, I can honestly say, putting up my own wind turbine is one of the coolest things I've ever done. I'm very disappointed in myself. I'm a huge failure as a feminist. I've let the team down. I love my solar panels, but when my wind turbine is wailing away in a high wind, I want to let out some crazed wolf howl at the top of my lungs, so that neighbors 4 miles down the road can hear me. I'm sorry, but I love my wind turbine at a cellular level.

In an off-grid set up, wind makes a lot of sense. The months with the least amount of sun, like November and December where I live, tend to be windy. The wind turbine complements the solar panels. Lots of off-grid systems therefore are a "hybrid," or a mix of solar panels and a wind turbine. This is a very synergistic, symbiotic relationship. In the summer when there's lots of sun, there is very little wind, so the turbine can be shut off. But once the days start getting shorter and cloudier and the cold November winds start blowing, the wind turbine is there to pick up the slack and help out.

Wind turbines have taught me a lot. The first thing they taught me is the "3 Month Rule". The "3 Month Rule" is something that a citidiot learns very quickly when they move to the country, and it's compounded when you move off the grid and you don't really have a clue about practical things, like electricity. The "3 Month Rule" simply states that regardless of how quickly you want to get something done, like fixing a wind turbine, it will take you 3 months, so you'd better learn some patience.

When we bought Sunflower Farm, Jean and Gary (the previous owners) had a wind turbine. Well, Gary had put up a wind turbine. Jean said it was "Gary's thing." In hindsight, with the masculine/feminine thing, it shouldn't have been a surprise it was Gary's thing. Gary had put up a turbine called a "Wind Baron". After the first energy shock of the 70's there were lots of wind turbines being produced. Many didn't have great designs, and there was a huge learning curve for the whole industry. The Wind Baron was one of these early designs and it was broken when we bought this place, but I wanted to fix it.

The turbine sat on a 60-foot tilt-up tower. The tower uses what's called a "gin pole". By having a piece of pipe perpendicular to the tower, you are able to distribute the forces when you bring the tower up, without it twisting or collapsing. You use a winch that pulls a wire that is attached to the top of the tower, and which is routed over the gin pole. Close to the winch you have a pulley that uses mechanical advantage to minimize stress on the tower and winch.

Gary had left the winch attached to the tower, but it was not mounted to anything. He had bolted it to his truck bumper when he needed to use it. I didn't trust my truck bumpers, and they didn't offer a convenient place to attach it to anyway. So I decided to cement it permanently in place, which meant that I could put the tower up and down conveniently. I dug a hole to cement in two steel pipes that I would affix the winch to. This time I was decadent and bought bags of ready mix which saved me borrowing a cement mixer and finding gravel and sand to use as aggregates. Without a cement mixer though, mixing it by hand in the wheelbarrow is a lot of work. You add water and mix, just like a cake recipe, but this is no piece of cake.

Once the cement had dried, I bolted a piece of 2 x 8 lumber to the pipes using U bolts. I put these on the pulley side of the pipe, or the side where the pressure was coming from. Ken, in his infinite wisdom took one look at it, and asked why didn't I bolt it on the other side. That way, as the winch was pulling the tower up, the force of the winch would be exerted against the pipes, rather than having the winch have a tendency to want to pull away from the pipes as I had mounted it. Ken even fabricated a piece of steel to bolt onto the wood, to add one more layer of strength to the system. If I could just be patient and see Ken's natural sense of logic when it comes to things mechanical, I would be a rich man.

Once I had the winch in place and ready to go, I was able to use a power drill to bring the tower down so we could get the Wind Baron off. The winch is an amazing piece of engineering, allowing you to move a lot of weight with just a hand drill. Once the Wind Baron was on the workbench in Ken's garage it was apparent why it didn't work. The turbine used a stator and a rotor with magnets, like a car alternator to produce electricity. As the wind moved the blades, the shaft turned magnets inside the wound copper coil to produce an electric charge. One of the magnets had broken off the shaft and had scraped up the copper windings, eventually jamming it. I could glue the magnet back on, but the copper windings would have to be re-wound.

This sounds easy enough, but as I learned, this alternator is wound differently from a car alternator, because the wind turbine does not move as fast as the alternator on a car. So it was off to Kingston to get the coil rewound. Of course, if you own an alternator shop, and you make your living fixing car alternators, you don't want to mess with a crazy wind turbine alternator that you've never worked on. So after a few tries, I brought it back home in defeat. I persevered though, and managed to find a shop that would take it on. Several weeks and several hundred dollars later, I had the rewound copper windings, the magnet glued back in place, and it was ready to go.

As usual, Ken suggested it might be worth testing first. Now there's an idea, test it before you haul it back up to the top of the tower. So we put it back on Ken's workbench and fabricated an adapter that allowed a drill to act like the wind and turn the rotor. And as I learned, this test was a good thing. It wasn't working. After several hours that day, and several other tries, wind turbines taught me a second lesson. Sometimes fixing an old piece of technology isn't worth it. Three months later I had no wind turbine and had invested lots of money and energy. So now what?

When in doubt, buy something new. If I was going to use the wind, I had better do it right. So I went about evaluating the commercial units on the market. Since we had moved off the grid, I had been reading "Homepower Magazine", which is the bible of do-it-yourself, home scale renewable energy (www.homepower.com). I had the August 2002 issue of Homepower with an excellent review of all the small wind turbines that were available. Mick Sagrillo, a wind expert who had been installing wind turbines for years had written the article.

Eventually I decided to install an Air 403 by Southwest Windpower of Flagstaff Arizona. It was a very aerodynamic looking unit that Southwest had been refining for years. They had sold more than 10,000 of the units and it came with a 3-year warranty, all of which sounded good. So I ordered my first Air 403 from George Wright of Metcalfe Wind Electric near Ottawa. When it arrived I convinced my friend Jerry Horak to come up from Burlington to help with the installation.

So on a windless morning we were ready to install the new turbine. Jerry and Ken set about attaching the turbine to the top of the tower, and connecting the wires from the new turbine, to the cables that ran to our battery room. This cable was extremely large, because the run was close to 300 feet from the batteries. There are two kinds of electricity. AC or alternating current, which is what your household appliances use. AC travels well over long distances, but you cannot store it. DC or direct current is the type of electricity that you can store in batteries. DC does not travel well over long distances, so Gary had purchased very large copper cable for this long run, otherwise much of the electricity generated by the wind turbine would have been lost.

Putting direct current electricity from small wind turbines into batteries is very common, because it's a fairly forgiving way to do it. The energy generated by your wind turbine will be quite inconsistent, varying with the wind, being very strong at some points and weak at others. Household appliances that use AC power require consistent electricity, 120 Volts and 60 Hertz. A wind turbine would produce "wild AC" which appliances would not like because the frequency would not be the consistent 60 Hertz they like. So it is better to have the wind turbine output DC and dump that into batteries. In an off-grid application this is good, because it allows you to store the power for when you need it, and then use your inverter to convert it from DC to AC for your household needs. The power that the inverter produces is of a consistent quality, better than you'll get from the power grid.

Even though it was my wind turbine, as is so often the case, I ended up being the tool fetch boy. This is a job I have had since I was a child. It involves being instructed on what tool is required, then sprinting the 1/3 kilometer to the house to retrieve it, and sprinting back. Although it sounds dramatically like Mel Gibson's role in his first movie "Gallipoli" where he plays a runner in trench warfare in World War I Turkey, it's not so glamorous. It was all right when I was 7 years old and it meant helping my dad, but now I was a man myself and sort of wished that I'd graduated to the tool dispatcher rather dispatchee. As I readily came to accept at the time though, Jerry, a skilled telecommunications engineer, and Ken, who is a licensed electrician and tradesperson, were infinitely more qualified to attach the turbine to the tower. I knew this would help me sleep when those big winds blew so I just shut up and ran.

Eventually the turbine was attached to the tower, the wires were connected and secured, and it was time to put it up. Jerry and Ken were going to take positions on the side guy wires where they would ensure that everything was going smoothly, and I began erecting the tower with the winch. The winch actually faced away from the tower, because the cable left the winch and went around a pulley that was cemented in the ground 10 feet away. From here it went to the top of the gin pole, and as I pulled in the wire, the gin pole pulled down which lifted the tower up. The pulley helped reduce the stress of pulling the wire directly from the top of the gin pole.

As you begin lifting a 60 foot steel tower with a 50-pound wind turbine, the greatest force is actually exerted just as the lift begins. The closer the tower is to the ground the more energy that's required to pull it up, and therefore the greater the force applied to all the components... the winch, the gin pole, and the pulley. Now the pulley seemed to be well cemented into the ground, but it appeared that maybe there wasn't enough cement. Knowing how much work is involved with mixing cement it would not have surprised me if Gary and his crew hadn't used enough of it, but unfortunately, the pulley was not the place to scrimp. As I continued to pull and the stress got higher, it appeared that the cement pad was actually pulling out of the ground. This was very disconcerting. If it did pull right out of the ground it was coming straight at me.

So I stopped, and shouted over to Ken, asking whether it was normal for the cement pad to be making a miraculous emergence from it's resting place, like a zombie from a horror movie. Ken made it very clear that I should stop, immediately. I guess this is where the whole male infatuation with all things wind comes from. As I had stood there watching the cement gradually being pulled out of the ground I had visions of it eventually being hurled through the air like a rock from some medieval catapult.

This seemed like an insurmountable problem to me, but it was a minor blip to Ken. Just bring the truck over and park it on the pad was his suggestion. With the weight of the engine on the pad, it would be fine. And it was. But I would be lying if I didn't admit to some mild anxiety once I resumed pulling the tower up. Luckily once you're past the half-

The winch cemented in place and bolted on to a steel plate. From the winch the wire heads to a pulley, now cemented in a pad, which includes reinforcing steel drilled into bedrock so it would not move.

way point, there is significantly less stress on the pulley, so I could relax a bit. I did highly regret having watched a war movie years ago where the resistance fighters had strung wires across a road at neck height as enemy motorcycles road past. If wires were to get snapping with the force of the tower, they would move faster than I could duck.

Once the tower was up there was a tremendous sense of accomplishment (or relief). Mother Nature was cooperating and the wind was picking up. Early wind turbines didn't come with anything as logical as a gauge to tell you how much power they were producing. For that I needed to make another series of phone calls to find a DC "Amp Meter" that would work. The manufacturer recommended an in-line fuse with the turbine as well, but they did not provide it. So off I went to Kingston again. This time I actually sat down with the President of a company called "Fusetek" to try and determine what I needed. This company had a warehouse full of fuses, but it was questionable whether they had one that would work for me.

This would seem pretty basic, but our Air 403 did not come with a way to turn it off. With any piece of electric equipment, the ability to turn it off is a nice thing. It allows you to stop it if there is a problem, or you need to fix it, or if something went wrong in the battery room and you didn't want the power flowing in. I guess the early wind engineers were focused on how to make power from the wind and keep the blades from flying off, so they didn't concern themselves with trivial stuff like on/off switches. But a switch is what they said I needed and so on another quest I went in search of a "50-amp single-pole double throw switch". Up until this time my exposure to DC electricity had involved the removal of DC batteries from my flashlight, and their replacement with new, comparable sized ones. I was extremely proficient at this.

I needed a DC on/off switch, which is quite common, but in most cases the wires coming into the switch are reasonably small. The 1 AWT wire running from our wind turbine was the size of large magic markers, so getting them attached to a small switch proved futile. The way that you turned off the Air 403 was by short-circuiting it, which means crossing the positive and negative wires. The wires I had to use were barely malleable, so the switch was becoming a real challenge. Ken came through as always, and fabricated an amazing clear plastic console. The console had room for the Amp meter which showed how much power the turbine was putting out. It also had the fuse which we had purchased, and Ken came up with an ingenious system for the switch. He simply had the positive and negative wires attached with exposed bolts. The positive lead from the wind turbine was terminated with a large alligator clamp, like you have on battery cables for jumpstarting a car. When the system was on, you kept the positive wire and clamp attached to the positive battery lead. When you wanted to short circuit it, you just took the clamp off the positive bolt and clamped it on the negative bolt and voila, it was turned off.

Many years later I learned this wasn't an optimal system because when you did this, it usually created a spark. When a large off-grid battery bank charges, it produces hydrogen gas. Hydrogen gas is explosive and we all remember the footage of the hydrogen-filled Hindenburg zeppelin exploding and burning. Luckily our battery room was well ventilated and the wind turbine never generated enough of a charge to get the batteries to gas off a significant amount of hydrogen.

The good news is that now, almost 10 years later, when you purchase many small wind turbines they will come with all the bits and pieces you'll need in a charge controller or power center like a fuse or circuit breaker, and an on/off switch

Our Air 403 was rated at 400 Watts in a 28 mile per hour wind. This

Nicole and Katie under the Air 403 on the old wind tower.

is an extremely strong wind. A "hold onto the nearest tree" sort of wind. We are not in an optimal location for wind, so it was rare that we saw a good strong consistent charge from this wind turbine, but there was the odd night we would get up in the morning after a windy night, and the batteries would actually be higher. This was a phenomenal experience.

The Air 403 had a characteristic that I loved. It was loud. Really loud. Insanely loud! I had always assumed that when you saw small planes with propellers that all the noise you were hearing was coming from the engine, but this wind turbine taught me that lots of the sound was coming from the propeller. The Air 403 used carbon fiberglass blades. In a high wind they would get very loud. With our nearest neighbor being 4 kilometers away, this was obviously not a problem. For me, it was the most beautiful sound I'd ever heard. It meant power. Oh sure, Mozart wrote some amazing music, as did Jimi Hendrix, but the wail of my Air 403 meant that my batteries were getting charged. It meant money in my pocket that I didn't have to spend on gas for the generator. It meant less greenhouse gases being produced at Sunflower Farm.

The 403 also had an even greater sound that it made. All small turbines have to have a way to stop or slow themselves at extremely high speeds. This ensures that they don't self-destruct in a hurricane. Many turbines have a moveable tail that furls, or moves the blades out of the wind when the unit reaches its maximum speed. The Air 403, as you can see from the photo, is one piece so the tail can't regulate its speed. It actually uses its blade design to slow it down. When the blades hit the top speed of 28 miles per hour, the blades actually twist to stall the system. This makes a horrific noise, like a transport truck gearing down a big hill, or a chainsaw on steroids. It was so loud, even if we were in the house, more than 300 feet away, watching television, we could hear it. I loved this sound. As Martha Stewart says, "It's a good thing."

Surrounded by woods, this was a truly awesome sound. Now if you were in an urban environment, that sound might not be so popular. We had first hand experience with this. Our friends Jerry and Ellen from Burlington had been coming up often since we had moved. Jerry who was constantly helping me with the system, started to get pretty enamored with renewable energy to the point where he decided to install it at his house in suburban Burlington. Now putting solar panels on a suburban roof is something I highly recommend. It's good for the planet and good for your piece of mind, as well as your pocketbook. But putting up a wind turbine in suburbia is another thing.

People in cities are used to noise of all sorts, but somehow a new noise, like that from an Air 403 is not something people have experienced before. Ellen and Jerry's house backed out onto the Niagara Escarpment and a 4-lane road that runs up it. Trucks going up the hill made a tremendous amount of noise as their engines gunned it to make it up. Trucks coming down often used their engines as a brake to slow them down, which was painfully loud at Ellen and Jerry's house. But for some reason the sound of their wind turbine was more annoying to their neighbors than the other sounds. Even though windy days tended to be noisy anyway with the wind rustling leaves and raising the level of background noise and even though most of their neighbors kept their windows closed to keep in the heat from the furnace or the cold from the air conditioner, they made it very clear that they weren't happy with the wind turbine noise.

Ultimately complaints to the city brought out the by-law enforcement officer. Using a decibel meter it was determined that the Air 403 was well within the acceptable limits. No louder than a leaf blower, or a lawn mower, or the neighborhood kid with a Honda Civic with a 5000-Watt power boosted stereo system. Just different. The moral being, that if you want to put up a wind turbine, make sure it's quiet, and try and get your neighbors on side.

North Americans have an interesting view of wind turbines. Perhaps it's because we missed the glamour of the European experience in Holland, but many of us don't like them. We protest when they are proposed in our locale. The best areas for wind turbines are often around large bodies of water, which is also where you find expensive homes and cottages. So even though the people with the houses along the water have kids with the same accelerating rates of asthma as everyone else, they'd rather have their power come from a coal plant, as long as it is somewhere else, even if the smog it produces still find its way to their child's lungs. We haven't come to accept wind turbines for the beautiful works of art and technology that they are. We'd still rather leave a legacy of 10,000 years of lethal nuclear waste than have to look at wind turbines. It's quite bizarre actually.

So we fight wind farms. We say they'll kill birds, and they're too loud, even though many protesters have never been near one. They do kill birds. A large wind turbine with environmentally sound siting might kill as many birds as a domestic cat, but no one is proposing that we ban domestic cats. The large buildings in metropolitan areas like Toronto often create huge navigational hazards for migrating birds. Pedestrians often have to step through the carcasses of thousands of dead birds in these situations, but no one is suggesting we tear down the buildings.

Standing at the base of a large wind turbine in a high wind is like standing beside the ocean, with the sound of waves hitting the beach. It is a pleasant, peaceful and hypnotic experience. After years of living with renewable energy, when I was finally able to stand at the base of a 1-megawatt turbine it was very much a spiritual experience. I was experiencing the solution to climate change. It was a profoundly moving thing.

Our Air 403 was not quiet, but it was good for the planet and I loved the noise it made. Until it stopped. Like a Don McLean song only in our case it was "The Day the Wind Turbine Died." Sometime in the second winter or spring after we had put it up, it stopped working. I'm not sure when. It's hard to tell. The turbine would still spin in the wind, but it would never spin fast enough to produce any power. Whether it was damaged in a windstorm, or struck by lightning, or had a defective part I was never able to determine. But that summer I took it down and replaced it with another Air 403. I did not go to the effort of shipping the broken one back for the warranty claim because we were supporting ourselves with our desktop publishing business, were editing a renewable energy magazine, trying to get gardens in, fixing things, and frankly, I just didn't have the time. I got the second unit at a very good price so down came the first unit and up went the second.

Putting the second unit up went much smoother obviously. Doing something for the second time usually does. And this time the winch was bolted onto its steel posts and I had replaced the concrete that held the pulley. In fact, in my third round of cement work, I was getting very fancy. One of the reasons the first concrete anchor had pulled out of the ground was that it was not very deep in the ground, because there was very little soil to put it in. The wind turbine was situated on a small hill 300 feet from the guesthouse. The hill was made of rock, with pockets of soil on it. So this time I dug until I hit bedrock. Since this wasn't going to give me enough holding strength, I rented a large concrete drill, and actually drilled 8-inch holes into the rock into which I placed reinforcing steel rod at various angles. This way, when the concrete hardened around it, it would stick tight to the bedrock and had nowhere to go. The drill was heavy and loud and created huge clouds of rocky dust. Anyone who fantasizes about living off the grid either needs a strong back or lots of money to pay someone else to do the work.

Into the cement I set a long piece of 4" square steel, which extended out about six inches. Into this I inserted a bolt that would hold the pulley in place. The bolt was very important. As the person who would be operating the winch, and be in the path of the wire if it were to break loose and fly back towards the tower at decapitating speed, I wanted this bolt to be strong. The bolt I used was of aircraft quality. I had learned about bolt strength while helping Ken fabricate his ultralight in his garage. Some people who like to fly purchase small ultralight planes. Ken chose to build his own. An ultralight consists of a small go-cart like seat and cage, which has a motor and large propeller, mounted behind the pilot. The cage is connected to a large wing, which is usually made of a set of aluminum ribs, covered in a strong vinyl material, like a wind surfer sail.

The cage where the pilot sits is connected to the wing with one bolt. Ken is a religious man, and as we were working on his ultralight, he presented a large, bronze-colored, nicely machined bolt to me and said, "This is the Jesus bolt." I was puzzled. What could a bolt that was the only thing keeping the pilot hanging under the wing, have to do with Jesus? As Ken said, "It's called the Jesus bolt, because if it breaks the only thing that's going to save you is Jesus!" Working with Ken on projects was a nonstop humorous learning experience.

So I had sourced my own Jesus bolt for the new pulley support, and this time I was much more comfortable lowering and then lifting the tower again with the new turbine attached. In our roles as book publishers we have spent a lot of time talking to dealers who sell renewable energy equipment. Many of them are not anxious to install wind turbines. There are lots of issues in terms of the installation; noise, bylaws, etc. But most of all, they are infinitely aware of the mechanical nature of wind turbines. When you put a piece of equipment up into the sky, exposed to the wind which can be very malicious and unforgiving you have to expect that sooner or later there's a very good chance something will break. The huge, million-dollar mega wind turbines are engineered to withstand these forces. The shareholders and business people who purchase them demand it. But at the small wind turbine level, the same does not always apply.

And so it came to pass, that within the three-year warranty period of our second Air 403, it too stopped working. When we examined the first one we took down, it was very difficult to see what exactly had gone wrong. There was no visible sign of damage as one might expect from a lightning strike. It was impossible to check the electronics because it is sealed in thick black poured plastic. I don't know whether this is done to protect it from the elements or prevent someone from copying their design but the end result was that it did not allow us to find out what the problem was. I have to assume that because it worked for the better part of two years, that the tower's grounding and the wiring were fine. But history has a tendency to repeat itself, and so it was when our second unit packed it in.

This time, serious action was required. I was still committed to using the wind to generate some of our electricity. The tower was located on a rise near the house, which offered a great view back towards the house and guesthouse. There were days, and some nights, when I would stand out on that rise, under the tower, and just marvel at it. There was a small red LED that would glow when it was producing power. It was easiest to see at night because it was 60 feet above me. Standing under that turbine, knowing that it was producing the power that was making the house glow with light, and knowing that no coal was being burned, no nuclear waste was being generated, and no natural gas (which s rapidly running out in North America) was being burned, was simply a wondrous thing. I was often overcome with a feeling of contentment. It seemed right. This was what I was meant to do. This was how I was destined to make my electricity. This is how people need to live in harmony with the planet. Using the resources like the sun and wind that are renewable. Yes, sometimes there are not enough of them, but this too is how humans must learn to live, within limits.

So for a while my dream of wind had to be put on hold. Work, gardening, and book publishing were pulling us in a dozen different directions. But then we learned about peak oil, and where the world stands in terms of how much readily accessible energy is left. I started growing concerned about being able to find gas to run a generator in the future. Would I be able to afford the gas or diesel to keep the lights on in November? I knew that I would be content to use the outhouse and go to bed at sundown, but I had a pretty good idea that many of my family members would not embrace these concepts. Having a fridge and a freezer to preserve our summer harvest from the garden was also something we desired. And so the dream of a new and better wind turbine started to grow.

As I read and spoke to others it was clear to me that I should be looking at a Bergey. Bill Kemp (author of *"The Renewable Energy Handbook"* that we publish) is one of North America's leading experts in renewable energy. He has a Bergey and it has been working for more than a decade. Mike Bergey had designed his turbines to be hurricane proof. They are commonly used on offshore drilling platforms and in remote locations like mountains to power remote telecommunications equipment. If I were going to go to the expense and effort of putting up another wind turbine, I wanted this one to work and keep on working.

I also decided that we needed to put this one on a tall tower. Wind turbines work best in open spaces, which is why the shorelines of large bodies of water are such a good place for wind turbines. Wind turbines don't work well around things that slow the wind down and cause turbulence. Things like barns and silos and trees. Living in the woods, we have lots and lots of trees. The trees had us surrounded. Now most of the time, this is a dream come true. But when it comes to producing wind power, being surrounded by trees would present a challenge. Our solution was to get the tower up above the tree line. The previous tower was 60 feet high and barely cleared the top of the trees. The new tower would be 100 feet high and would clear the trees by at least 30 feet, which is the recommended height above sources of turbulence. Even on a 100-foot tower it was still not going to be a perfect location, but it was going to be the best we could do. In those windy months like November and December we were going to displace generator run time, and that's what mattered.

So I started to hatch my plan for the new system. Anytime I was over near the old wind turbine I was thinking about the set up. There is a rough road in that direction that I use to take the pickup truck in to bring back firewood I've cut, but getting a fully loaded cement truck down it was going to be another story. I'd have to do it in July or August, during a drought, because the road turned into a quagmire when it rained, and the last thing I wanted was a cement truck stuck down that road. Then there was the issue of getting a truck into the areas where I need the cement. I know what you're thinking. Why didn't I just mix it all by hand like I'd done in the past? Well, after 8 or 9 years of living here and numerous projects that had involved mixing concrete by hand, I had come to realize that the volume of concrete I needed for this project did not lend itself to hand mixing. This was a tall tower on poor quality soil, so there was going to be a lot of concrete holding the anchors in place. I'd learned about what happens when you don't use enough. It has a tendency to want to leave the hole in the ground where it was placed, and frankly, that is not a good thing.

This hill was going to make getting the truck in very difficult. Eventually I had sort of resigned myself to having to fill one of the anchor holes with concrete that we wheeled to it in wheelbarrows. For the third hole location, I'd have to get the cement truck to dump cement into our wheelbarrows, and I'd have to hope I could find some strong, enthusiastic friends or neighbors to help. This was not a one-person job. I'm sure they'd approach the request with the same enthusiasm I greet people with who ask me to help them with a move. You know the ones. They live on the third floor of an old house and they have narrow stairs and lots of big bulky furniture. And they're moving to a fourth floor walk up. And they're doing it in August. During a heat wave. Sure, I'd love to help!

I called Michelle over one day to show her where I had staked out the new anchors, and where the tower would lie down before we pulled it up. I mentioned my concerns about cement trucks getting stuck, about the additional work in hauling cement in wheelbarrows, and about the hassles with having the tower so far from the house. It doesn't matter how much you prepare, you will need 10 trips an hour back to the garage to get the tool/tape measure/bolt/shovel/wheelbarrow/level/cordless drill that you forgot to bring. Oh sure, that's a fun walk the first couple of times, but it gets old fast.

On the walk back as we got close to the house Michelle said "Why don't you put it here?" My response was "Because that's where the wind tower goes, over there." "Why?" "Because that's where it's always been." "Why?" "I don't know, that's just the way it is and you should accept it." Michelle's questions made me think, and I wondered, "Why did I have to have the wind tower 300 feet from the house?" The wire was in place, but I could move the wire or buy new wire. Gary had picked that spot because it was on a bit of a hill, but it wasn't more than 5 or 10 feet higher than the elevation around the house. With a hundred foot tower, that wasn't going to be a big deal.

As always, Michelle had had a stroke of genius which was going to save me a ton of work and was simply a much more logical place to put the wind turbine. Part of the area that we chose was already cleared and the rest had some small trees, mostly sumacs, so I had to start clearing them. Then I had to find the best location for the base of the tower. The four anchors had to be located 50 feet from the base. I had decided to make sure that the base of the tower was also more than 100 feet from the house, guesthouse, solar panels or anything of value. Yes I was going to be erecting this tower to last a lifetime, but one still has to factor in "catastrophic failure" when positioning such a structure. My daughter Katie, whose bedroom is situated on the same side of the house as the wind turbine, was relieved to hear this part of the equation.

Once I had the base of the 100-foot tower located 105 feet from the house, I could conveniently locate three of the four anchors. Unfortunately the fourth location was in a small depression and was 4 or 5 feet below the grade of the other anchors. This was the sort of hole I wasn't going to fill up with a shovel and wheelbarrow. So I called out the heavy artillery and hired a backhoe to come and fill it in. There was lots of sand in the area, because it was beside the pond, which we had dug out when we first moved in. The backhoe previously had just put the sand it dug to both sides of the pond, so there was lots of sand available to fill up the depression. While I had the backhoe here I ticked off a whole list of other jobs that I had him do as well.

Next I had him dig the holes for the anchors. We sit on the sand and rock of the Canadian Shield; a massive area of igneous rock than eons ago was molten lava. Sometimes the rock shows through the soil, and sometimes you can dig great distances and never hit rock. It's a crapshoot. The sand that sits on the rock is just that, sand. Sandy sand. Like you see at the beach that you can make sand castles with. When there is some moisture in our sand, it will hold together. But when it's dry, it has no strength or structure. As our climate changes, our area is going to be very prone to drought, because it already is. When the backhoe began digging the holes for the anchors, we hadn't had rain for a month. So while the backhoe was making fast work of digging the holes, nature was making fast work of filling them back in as the sand on the sides collapsed into the hole.

When he was finished the first hole it looked much bigger than I imagined. And with the way they kept caving in, it looked like I was going to have to shore up the sides with wood, to keep the hole to the size I wanted. Concrete is expensive, so I didn't want to order any more than I needed. So I made the fateful decision to dig the holes by hand with a shovel, and I sent the backhoe off to work on other projects, like clearing the stumps from around the barn foundation where I wanted to expand the garden. Digging a hole in sand is no big deal, within reason. My holes were going to be about one and half cubic meters, or about 3 feet by 3 feet by 5 feet deep. Once you get the sod off the top it goes very well. The first foot or two is a walk in the park. Even the third foot is not too bad, if your hole is big enough and you can stand in the hole and toss the sand out. Once you get past that third foot it's a different story though. Now you have to lift the sand up past your waist, and if you're using a regular shovel, the end of it will constantly be hitting the wall behind you, so you sort of have to maneuver the shovel up and out. Then when you get past that fourth foot, you're into no-man's land. I'm 5 feet 8 inches tall, so when I'm digging a hole five feet deep (to get below the frost line) I had to lift and toss the dirt almost over my head. As the holes got deeper I couldn't help but think once again "I've made a terrible mistake!" It wasn't just that I had the 3 other anchor holes still to do. It also was a 4th hole for the anchor, which the winch would attach to for pulling down the gin pole, and then a 5th for the tower base. I required 6 holes in all.

This was not a situation I could get out of quickly. It was hot, so it was just a matter of digging early in the morning or later in the afternoon, and then doing lots of stretching to compensate for how my body was rebelling against the whole hole-digging regimen. I had dug ditches to lay power cables to the barn, for Ethernet cables to the guesthouse, and for power cables down the sidewalk to the guesthouse for outside lights, but they were 2 feet deep. This 5 feet deep thing was a challenge to be reckoned with. In hindsight I should have just had the backhoe do it and figured out a way to use plywood or particleboards to shore up the edges. Live and learn though. Now my philosophy has become as long as there's diesel fuel available for machines like backhoes, I'm going to use it.

Eventually the holes got dug. Now I had to get serious. I had decided the Bergey was going up in September. I had noticed a theme in articles in "*Homepower Magazine*" and on websites about people putting up wind turbines. They inevitably seemed to be working in the snow. I think it's because the magnitude of the task escapes most people, so they start working on it, maybe scoping it out in the spring. Once they've decided where the turbine is going they need to think about digging the holes. A call to the local backhoe operator informs them that it's the busy season, so there's a delay until the backhoe can come. A month or two later the holes are dug, then they realize that they need reinforcing rod, then when the tower arrives they discover that they are missing a key piece, then when they try to connect the turbine to the tower they realize that 3/4" bushing yattity yattity yattity is broken, which has to be reordered, which is out of stock at the supplier, and the next thing you know, it's snowing.

That was not going to happen to me. The key to the whole process was going to be getting the anchors cemented in the holes, and the concrete cured. We were going to be taking our first vacation away from the house the third week in August, so the concrete was going to be poured by then, come hell or high water.

When you fill up a hole with concrete, you have to give it a skeleton to cling to and give it strength, otherwise when it freezes and weathers, it will crack and separate and weaken. I had decided to build a reinforcing steel cage that I would suspend in the hole while the concrete was poured in. I did not want it to sit on the bottom of the hole, because the ends of the steel would be exposed to water and would have a tendency to rust and draw that rust into the anchor. You often see this on highway overpasses where construction crews are often seen replacing the rusty steel rods. Once I had purchased the steel, Michelle and I went over to Ken's garage to make up the cages. This involved bending each piece of reinforcing rod twice, to form a big "U" shape, then attaching four of them together. First we'd hold them together with plastic tie wraps, and then we'd use a piece of wire and tightly twist it to tie each piece to the others.

Once the reinforcing rod "baskets" were made up, we took them home and suspended them inside the holes. By that time we had picked up the tilt-up tower and turbine from Renewable Energy of Plum Hollow. Even though they don't install wind turbines, they reluctantly agreed to order one for us. The tower is made of 10 sections of 4" tubular galvanized steel, plus the gin pole, plus the guy wire kits, plus the anchors. It was 1200 pounds worth of equipment. Our 1993 Ford Ranger pickup is a small pickup and never had good springs, so this was clearly beyond its capability. So we convinced Ken to help us. He and Michelle drove to

One of the anchor holes with the suspended wire cage in place to add strength to the concrete. The plastic liner is to prevent the dry sand from wicking moisture from the fresh concrete, causing it to dry too quickly.

Kingston in his truck towing the trailer that he used for his ultralight. That afternoon as I was expecting them back, I heard the truck pull in and went out to greet them. There was no trailer, which concerned me, and Ken looked a little "focused."

"Have you got some steel, and Ubolts, and…?" Apparently the weight of the tower was even too much for his trailer. The trailer was built to hold an ultralight plane with the weight distributed over the axle. The steel pipe had placed too much weight toward the front of the trailer and it had eventually broken a weld. I was just grateful it had happened 5 minutes from our house on our quiet road rather than on a 4-lane highway. It was raining lightly as we gathered up gear, but by the time we got to the trailer, it was pouring. Not just raining heavily, but a "fire and brimstone, time to build an ark" deluge. Ken and I sat and waited it out. It would have to end soon. It never stays this heavy for very long. Five minutes passed. Ten minutes passed. Finally we said "Damn the torpedoes" and we got it ready to go, while getting drenched by the downpour. Soon after, the tower and turbine were in place in the yard close to where they ultimately were going to be erected.

As part of our book publishing business, we have branched into DVD production. First we shot and edited a companion biodiesel DVD to go with our "*Biodiesel Basics and Beyond*" book. Then the following summer with our daughter Katie home from University, we had her film a gardening DVD, which I narrated and which was a basic introduction to turning your backyard into a produce department. So with us putting up a new wind turbine, it was natural that our next video project would be how to put up a small, home-sized wind turbine. While our renewable energy book was an excellent resource, I found the Bergey manual confusing in some places. Some people are visual learners, so I think being able to watch someone going through the various tasks involved with erecting a tower and wind turbine is very helpful for a lot of people. In fact, I wish I could have watched it myself before I started.

So not only did I face all of the challenges of putting up a wind turbine such as digging holes, making reinforcing rod cages, cementing anchors into place and all the other related activities inherent in a project like this, I had to make sure that I filmed everything. When we produced our gardening DVD we used a small video camera that fit on a typical home-style tripod. It produced excellent quality video, and was very portable. Now that we were going into video in a big way, we had upgraded our camera to a much larger and heavier one, which required a much larger, and heavy, professional grade tripod. It just added one more dimension to the project. I had to figure out what had to be done, pick a time when the light was good, take all the tools I needed, take out and set up the tripod, take out and set up the camera, get Michelle to film whatever was being done, and often, take a few takes to get it right. Since I wasn't scripting things, and was often trying to put things I'd never put together before, I would usually try it, mess around until I figured how to do it like I knew what I was doing, then film it like this was second nature to me. Whoops, hope I'm not giving away trade secrets.

I think you can tell from the DVD that these are real people doing the work. The hole I dug nearest the pond was a favourite place for frogs to tumble into. So one of the scenes has me tossing out frogs (gently of course) before I filmed the sequence, getting the hole cleared of all the unpaid extras as it were.

To suspend the reinforcing rod cages in the holes we put two pieces of T-steel, which the local feed mill sells for erecting fences. We laid them across the hole, and then hung the cages underneath with coat hangers. These rested on the wooden frame I had built for each hole, which was set down into the ground on top of the hole. These made an even guide for the concrete to flow into at the top of the hole; otherwise it would be

Ken helping with pouring concrete into the anchor holes.

very jagged and more likely to crack around the edges. We also suspended the anchors that came from the manufacturer so that the anchor pointed toward the top of the tower at about a 45-degree angle and extended about 18" out of the concrete.

Getting the cement to the new location that Michelle had suggested was going to be much easier than to the old wind turbine location, but there was one challenge. The heavy cement truck was going to have to drive over the new sidewalk I had made that runs from the house to the guesthouse. I loved this sidewalk and we used it constantly and if it cracked and got uneven, it would break my heart. So we got a load of topsoil dumped on the sidewalk. It was truly one of the craziest things I've ever done, but that's all relative. We spread the topsoil out, and drove our truck over it to pack it down. The object was to distribute the massive weight of the cement truck over a greater area. When the cement truck was done and we removed the topsoil, it had worked!

The concrete pour went well and as usual Ken finished off the cement as if he was in the foyer of a grand cultural hall. We let it cure for a week before we started the next phase. I laid out the sections of the tower and inserted the wire through each section. I then got the backhoe back to dig the trench from the base of the tower to the battery room. I had learned my lesson from digging the anchor holes. The backhoe dug the trench in half an hour, we threaded the wires through conduit and laid the conduit in the trench, and it was backfilled within the hour. When I saw how many tree roots the backhoe went through, I know it would have taken me a month to dig the job manually.

Now we were getting very close. I had mentioned the project to Steve Lapp, who runs the Renewable Energy Technologists course at our local community college, St. Lawrence College. Steve was interested in having students come up and help with the erection as practical work experience. So we set a date 3 weeks off and I scrambled to get everything finished in preparation. As we got closer to the date, the 6-week drought continued, until the night before the tower was supposed to go up. As can be expected, the drought was supposed to end September 28th, the day our tower went up. The forecast suggested a 70% chance of rain. And so I resigned myself to it being wet the day I put up the tower. It was not going to help with the filming.

Luck smiled on us the next day though. It was sunny in the morning as the 15 students arrived. Steve was thorough in instructing them as to the proper way of completing each task. Ken found this frustrating because he's a "let's get this show on the road" kinda guy, who deals with stuff as it comes up. Having the skills Ken has, he can get away with that. Teaching students the proper way of putting up a wind turbine though, was much more methodical.

You actually put the tower up twice. The first time you put it up without the turbine on, so you can set and tighten the guy wires. Then you lower the tower and attach the wind turbine and pull it up again. Attaching the wind turbine provided lots of learning opportunities for the students, like how to torque down a bolt to the correct amount. You don't want your blades sailing off into the sunset, so when the manual says, "They should be torqued to 40 ft-lbs", that's what you do. When I realized that some of the students had never used a torque wrench before, I had one of my "I've made a terrible mistake" moments. Luckily Ken was there watching and guiding. Ken's talents were wasted working in a prison all his life. He is a natural-born teacher.

The whole process had taken much longer than anticipated. It was now 3 p.m. and the sun, which had been so great in the morning, was gone. In its place were very threatening-looking storm clouds. Naturally, when the choice is, do we try and get this wind turbine up before the storm hits, and the question is being asked of 15 males, the obvious an-

Ken, third from the left (without the hardhat) coaches students on how to correctly use a torque wrench to fasten the wind turbine blades.

swer is "Let's do this!" So using my 18V cordless drill, we started pulling up the tower. The clouds were getting darker, and the wind was picking up, but really, if nature wasn't threatening, how much fun could it be? And where else do you want to be working when the lightening starts but standing under a 100-foot lightning rod which screams "Hit me, hit me!" at the storm? And just like something in a scripted Hollywood movie, the tower was up just as the rain hit. There was lightning and thunder and I insisted, "Everyone step away from the tower."

So we retired to the garage to watch this wonderful new machine. After the crew had left I spent some time cleaning up tools and organizing stuff. The rain cleared, the sun came out, and I could not drag myself away from this machine. It was after 6 p.m., I hadn't slept well the night before, had been up at the crack of dawn, had run all day, could hardly walk I was so tired, but I simply could not take my eyes off my new wind turbine. It had taken more than three months to put it up. It had been a huge challenge for me. Every part of it, from anchors, to assessing wire, to positioning the winch, had been outside of my realm of experience or comfort zone. But I wanted this machine up and I was going to do it myself so I knew that it was properly installed and so I knew how to take it down to get it fixed if it broke. I have no womb, but I felt like I'd just given birth – to a wind turbine. I knew what it must be like for someone who has just designed a skyscraper, or built a plane that flies for the first time. As much as I am drawn to nature and simple things, putting up this 100 foot piece of engineered steel is simply the coolest thing I have ever done. If I were ever on the borderline of losing my attraction to solar panels (which I'm not), my wind turbine pushed me way over the edge.

As my karma with so many things mechanical is, within a few months of the turbine being up I noticed it was behaving strangely. It was starting up in extremely light wind and spinning far too fast, basically free wheeling as if no loads were on it. When it was working properly it needed fairly significant wind to start and didn't free wheel. A check of the Amp Meter told me that no electricity was being produced when it was behaving like this. Sometimes it worked and sometimes it didn't. So thankfully I'd used a gin pole tower because it meant I could pull the tower down conveniently. Ken and my neighbor Sandy helped me bring it down. One person operated the winch, powered by my Dewalt 18Volt Cordless Drill and one person on each side anchor to check the wires as we brought it down.

My decision to buy a Bergey was based on its reputation for being a rugged, American engineered and made piece of equipment. Of course, with the nature of capitalism and competition when the box arrived it had a "Made in China" stamp on it. And when we took the turbine apart we could see the problem. The brushes were not working properly. The brushes are actually pieces of copper that sit in a groove and as the turbine spins and turns they allow the electricity to pass to cables that are fixed in the tower. The brushes seemed to be too small for their running groove and you could see they were actually turning on their side at which point they weren't making contact. When I pointed this out to Bill who had an earlier version of the Bergey he was surprised at the new design which wasn't as robust as his.

Now Bergey was happy to replace the brushes with better quality ones, but it's one of those things you'd rather not do. They insisted I was the only customer this had ever happened to, but this just seemed to be a design flaw. And it's why you need to approach wind with caution. When you install solar panels they just sit there and work. A wind turbine is a mechanical device that exists in a very hostile, high wind, environment. So there are much more likely to have problems. Our dealer in Kingston,

I use my DeWalt cordless drill to raise and lower the wind tower.

Renewable Energy of Plum Hollow doesn't even like to sell wind turbines anymore. I had to beg them to sell me one and they told me I was on my own with it. I guess if I were a dealer trying to earn an income having to fix wind turbines, which break, versus just installing and never having to touch a solar panel again, I'd be doing the same thing.

The whole exercise was a good experience though. It showed me how to get the tower down with minimal help. I could technically get it down by myself if a hurricane was forecast to blow through. The turbine is designed to furl and deal with that high wind, but at least I know I have the option.

And I also learned about how great Dewalt cordless drills are. So great in fact that Ken was constantly borrowing mine for jobs he was working on. So I got smart, and bought a new drill and gave Ken my old one. Each drill comes with two batteries. It takes the charge on two batteries to get the tower down, but it takes 4 batteries to get it up. So now I had access to 4 batteries when I need them. And with someone like Ken who has been instrumental in so many components of our life off grid, it's the least I could do. And yes, I was greedy and kept the new drill for myself.

As I sit writing this book, my desk faces the wind turbine, and the wind turbine is positioned perfectly so that I can see it from the bottom

to the top. It's as if I had planned it in such a way as to be visible from my desk chair. But its location was just a fluke. Knowing how many great coincidences we've had here though, it does not surprise me. It was destined to be positioned so well. Putting up a wind turbine can be a huge amount of work, but it was worth every minute of it!

I love my solar domestic hot water heater. I love my solar panels. But I REALLY love my wind turbine!

Eureka! Houston we have wind power!

It's better than any reality TV show to watch my Amp meter bouncing up and down on a windy winter night as my batteries are charged, even after the sun goes down.

10 Realities of Country Life

by Michelle

I was raised in the suburbs. As a child I walked to my local elementary school and later to my local high school. As an adult living in the downtown core of Burlington I was able to walk or ride my bike to almost anywhere that I needed to go. I had a choice of two grocery stores within walking distance. The library and my daughters' elementary school were both a walk away. The hair salon where I got my hair cut, my doctor's office, my dentist's office and even City Hall were just a short walk from my home.

What a culture shock it was to move to the country. As I've mentioned, our nearest neighbors are 4 kilometers (2-1/2 miles) away. If I want to borrow a cup of sugar it means a 4-kilometer walk or bike ride or more likely a car ride. Shops and services are even further away. The village of Tamworth is 14 kms (8-1/2 miles) away on a very unpopulated road that is at times curvy and hilly. A bike ride to town is not a journey to be taken lightly!

When we first moved here, there wasn't a grocery store in town. So I moved from one home where I was able to walk to my choice of grocery stores, to this home in the country where the nearest grocery store was a half hour drive away! In fact, it was a trip to a grocery store shortly after moving here that provided me with the first "culture shock" experience. As we prepared to move to the country we decided we would need a second vehicle. We had survived quite easily with just one vehicle in the city but we didn't want one of us to be isolated without a vehicle. So we purchased our new neighbor Alyce's used truck. On this particular day I drove the new-to-me truck to a little town about ½ an hour away. I did my shopping and came out of the store. First of all I had to remember which vehicle that I had driven, and I looked around the parking lot for the sight of a truck that I hoped would look familiar to me. Once I found the truck I sat behind the wheel and had to give some thought as to how to get home!

When we lived in the city I rarely drove our vehicle. Consequently I rarely had to pump gas. Remembering to check the level of gas in my vehicle and pumping gas when necessary were two other skills that I learned pretty quickly after moving to the country.

When we moved here we were "home schooling" our daughters. They were 12 and 10 years old at the time and had been home schooled for 4 years. Unfortunately we quickly learned how much harder it is to find other home schooling families when you live in a rural area and how much harder it is to get together with them. Once again I found myself driving a lot more than I was used to in our efforts to provide social interaction for our daughters.

After a couple of years of a rather isolated home schooling experience my daughters were approaching the age at which most children head off to high school. The lure of daily social interaction was very strong and they both decided that they'd like to attend our local high school. "Local" in the country is a relative term and our nearest high school was located about 48 kms from our home. We drove down and arranged a meeting with the principal and he welcomed my daughters as new students to his school. I called the local bus company and provided our address and was told that a school bus would pick my daughters up out front of our house at 6:40 a.m. on the first day of school. For two girls who were accustomed to a more leisurely wake up routine, this was going to be a shock to their systems! Being the thoughtful mother that I am, it was going to be a shock to my system too, getting up early enough to wake them and help them to get ready for school.

Sure enough the empty school bus arrived early on the appointed morning. The driver lived in town and so that is the direction the bus came from. The driver asked if he would be able to turn around in our driveway. I gave him permission to do so but he was reluctant to attempt it, given how narrow our driveway is with a dense forest on both sides.

He ended up driving past our house and down the road in search of an appropriate place to turn his bus around. He drove 2 kms (over a mile) before finding the next roadway and immediately I felt guilty about this empty bus traveling not only 14 kms from town but also an extra 4 kms (2-1/2 miles) so that it could be turned around safely.

That was just the beginning! The girls hopped on the bus and were driven by the first bus driver to a corner closer to town. They got off the first bus there and got on to a second bus. This bus took them out to the highway where they again had to change buses to the third and final bus, which took them the rest of the way to their high school. They arrived at school slightly less than 2 hours after they had left home!

Luckily the journey home wasn't quite so bad and required only two buses! Even still they left their school at 2:30 and the bus didn't deliver them to our door until 4:15 or so.

Cam and I were not only unhappy with the timing of these multiple buses, but we felt it wasn't right for an entire empty bus to drive all the way out to our place to pick our girls up. So we began negotiating with the local school board to pay us mileage to drive the girls in to town, since they were basically the only students on the bus. We were told that we didn't qualify, but Cam kept after them. They bent the rules numerous times for other students who were on dead end roads that buses couldn't turn around in, or had access limitations. We reasoned with them that the carbon footprint of driving our Honda Civic to town, rather than a huge, empty school bus, simply made sense. Eventually they capitulated after we wore them down with our persistence and they started paying us mileage to run the girls into town.

So we began driving them into town where they were able to get picked up by Bus #2 and only had to transfer once on the highway. We also began picking them up in town at the end of the day so they only rode in one bus on the way home. This shortened their day considerably, allowing all of us to sleep in a few minutes longer in the morning and also got them home earlier in the afternoon. Since we don't receive mail delivery out here, picking the girls up in the afternoon allowed us to pick up our mail and any groceries we needed at the same time.

Aside from all of the downsides of living in the city, urban life sure has some positives, some of which you don't recognize until you've moved away. In the city, when we had one of those nights when neither of us felt like cooking (but we still felt like eating) we just picked up the phone to order a pizza or Chinese food. This is not an option in the country. Sure, I can hop in the car, drive 14 kms and buy a pizza from our local pizzeria, but all of that effort defeats the whole purpose of being able to buy a ready-to-eat dinner.

Pizza became one of the first previously purchased meals that I learned to make on my own. Today I can whip up a pizza crust and top it and have it in the oven within about 30 minutes.

When your nearest grocery store is a 30 minute drive away, you learn to stock up on basics. We are very fond of bread. In the city I had my choice of bakeries where I could purchase a variety of breads and bagels and other treats. Cam and I have always preferred whole grain breads preferably with lots of nuts and seeds in them. Without an actual grocery store in Tamworth, we checked out the local corner store in town, which stocked bread and milk and snacks. White Wonder Bread was pretty much the only bread available. There is a bakery in town, but it closed in December and didn't reopen until April. So I was faced with the prospect of a half hour drive every time I needed a loaf of bread, or I needed to learn to make my own. Luckily I had never been averse to baking with yeast and so I just began making homemade bread more often.

We have no mail delivery to our rural home. Instead we have a post office box in town and so if we want to mail something or check our mailbox it means a drive into town. We don't go in to town every day and so sometimes our mail piles up. The advantage of this postal situation is that we get to see and converse with our local postmistress a few times a week. In the time we've lived here, we've discovered what an incredible source of information post office employees can be. We are lucky to have Katie as our postmistress – the friendliest and most helpful friend you'd ever want!

We don't have garbage pick up here either. In the city, once a week our garbage and recyclables were picked up at the end of our driveway. Now we let these materials pile up in our garage and when they are finally threatening to take over too much space out there, we load up the car or the truck and take them to the dump. This system has a few benefits. First of all, it encourages us to make as little garbage as possible. Even when we lived in the city we were very cognizant of how much garbage we were producing. By reducing our purchases, reusing things, composting and recycling, we were able to limit our garbage to putting out just one can four times a year. In fact during Waste Reduction Week the local papers and TV station came to interview us to ask how we had managed to limit our garbage. Our neighbors often put out 2 cans EVERY week! Now that we are faced with the prospect of a 15 minute drive to the dump, we make every effort to limit our garbage output even further. We are usually able to go 3 or 4 months before we finally have to take our 1 or 2 bags of garbage and boxes of recycling to the dump/recycling depot.

Along with our lack of postal service and garbage pick up we also discovered pretty early on that many trades people aren't too fond of coming out this far to work. When Cam and I moved here, we were not the least bit "handy." If our taps started dripping in the city, we called a plumber. If we wanted some wiring done, we called an electrician. We just assumed that we would continue to use the services of trades people in our new home in the country. We soon learned that trades people could be very selective in returning phone calls. Too often we would call and leave a detailed message, telling the plumber or the electrician what we needed done. In the city we could often count on seeing them the same day. Here in the country we learned to wait a day or two or more before our phone message would even be answered. A promised visit to come out and take a look might take a few more days. Some times the promised visit just never happened. Pretty quickly Cam and I discovered that it was just much easier to do the job ourselves. Luckily we are blessed with a wonderful and talented neighbor and every time we have needed to learn a new skill in order to make a repair, Ken has been there to lend a hand and show us the ropes.

Of course sometimes Ken's incredible confidence in his skills (and in ours) have gotten a little out of hand. Our front sidewalk is a case in point. When we moved in, a very charming and picturesque fieldstone pathway joined the house and the guesthouse. The fieldstone looked lovely. The problem was that over time the frost had caused the rocks to heave and shift and the pathway became more and more treacherous. This problem was compounded every spring when the snow on the pathway would melt during warm days and then refreeze at night. A trek between the two buildings was downright dangerous. When Cam mentioned to Ken that we were considering replacing the fieldstone pathway with a cement walkway Ken was incredibly supportive. He assured us that this was a weekend job and that he would help.

So Ken helped Cam to scope out the job and the project was underway. First of all, Cam had to remove all of the old rocks. They had settled over time and were now firmly held by the surrounding soil. It was a huge job digging each of them up and removing them and piling them off to the side of the lawn. Cam was able to do this job a little bit at a time but it was a major undertaking. Just this preliminary step took him about two weeks.

Once the pathway was cleared of rocks, Cam needed to cut the sod to make a reasonably straight and level pathway of a consistent width between the buildings. He needed to lay down a layer of sand and level it out the whole length. He also needed to place some reinforcing steel mesh and then construct wooden forms for the cement to be poured in to.

Finally all of this preliminary work was done. Needless to say, it took much longer than just one weekend. He called in a cement truck and arranged to have Ken and some of his relatives to help pour and deal with the wet cement. The job went well and we were rewarded with a nicely raised, level, straight sidewalk allows rainwater and melting snow to drain. Cam is usually able to keep the sidewalk cleared of snow and ice all winter, allowing us to move between the buildings without putting waterproof boots on! We've learned that when Ken says that a job will take "a weekend" we'd better adjust that estimation up... by quite a bit! In the city you can pay people with skills to do what needs to be done. Here in the country we've learned that it makes more sense to acquire those skills yourself!

I had almost forgotten to mention this last challenge of country life, until one of my friends who read over this manuscript asked "aren't you going to mention the bugs?" I guess after living here for as long as I have I have come to accept each of the pests that accompany the changing of the seasons around here. As April showers taper off and the nice weather of May arrives, so do the mosquitoes and the blackflies. Cam often wears a net hood when he gardens or works outside during this time of the year. I prefer to just put up with them for as long as I can before they send me running for cover! As the days heat up, the mosquitoes and blackflies confine their activities to dusk and dawn. Then the deer flies arrive. They swarm you as you walk around on hot, sunny days but I've found that of you stay still, they leave you alone. Cam wears a sticky patch on the back of his gardening hat to trap as many of them as possible.

On the plus side there seems to be fewer fleas out here in the country. When we lived in the city, our backyard was full of squirrels, and I've been told that they are notarious for being flea infested. We always had problems with fleas on our cats in the city. I haven't seen a flea on our dog or our cats in the whole time we've lived in the country.

Despite all of the challenges of living in the country we continue to enjoy living here and find it increasingly difficult to spend any time in the city. Cities now seem much too crowded and busy and noisy for my sensibilities and I find it exhausting to spend much time in one. Trying to sleep in a city is even harder – the bright lights and constant noise make for very restless sleeps for someone who is used to the total darkness and absolute silence of life in the country.

11 The Big Blue Monster in the Garage

By Cam

Living off the grid has challenges. It has many challenges. It is incredibly gratifying, but this does not come without some downsides. One of the really eye-opening concepts of off-grid living is the idea of things being "finite". When you live connected to a large, centralized electricity grid, where someone else provides electricity to you, one starts to think of electricity as infinite. It doesn't matter how much energy you use, how long the television is on, or how cold it gets outside while your heater is trying to keep your hot tub on the back deck at 120 degrees in the middle of winter, the electrons will continue to flow through the cable to your home.

When you live off the grid, you quickly become aware that electricity is finite. It has a set, clearly defined end point. If the sun doesn't shine, and the wind doesn't blow, sooner or later, your batteries will run out of energy, and you'll be left in the dark. While some intrepid off-gridders do in fact live this way, we are not so inclined and wish to have electricity available to us, all the time. So when the battery meter is starting to approach "empty" it comes time to set aside our environmental ethics and run a fossil fuel power generator. It's not something we do lightly, but if we want to continue to operate our home-based business, and keep the lights on and the water flowing from our well to flush our toilet, and keep the food in the fridge and freezer from spoiling, a generator is really the only solution.

Generators are the "dirty little secret" of off grid living. Lots of off-gridders don't like to talk about it, but if you're going to live a typical North American lifestyle, you need a generator. How much you run it depends on how you approach climate change and the concept of peak oil. Lots of people move "off grid" and "on to propane", which means shifting all of their major thermal, or heat loads, to propane. This is sort of cheating if you're doing it for environmental reasons, since propane is a non-renewable fossil fuel. Other people move "off grid" and "on generator," continuing to live as they did before, just making up the difference that their solar panels don't provide with a generator. We are working to make sure we're not in either of those camps and we hope to eventually pull the plug on all fossil fuels and live "zero-carbon."

Ironically enough, not only are generators the part of off-grid living that we are least comfortable with; they have also been our greatest source of problems. Sort of poetic justice I think, being punished by the god of carbon emissions. Usually I try and stay positive and use the word "challenge," but our generator saga is beyond a challenge, it's been a problem, plain and simple, bordering on a pain in the butt.

The generator that Jean and Gary left when we purchased the house is an Onan, 10,000-watt, gas powered one. It's 40 or 50 years old, and is quite an amazing machine. It has many hours on it and continues to churn away. Gary had purchased it from a local farmer who had it stored in a barn. It had sat for many years and nature had started to reclaim it, as a number of creatures had made homes living in it, including snakes. Gary took it to a shop in Kingston, where they restored it, and got it working.

The nice thing about the Onan was the fact that it runs at 1,800 rpm, rather than 3,600 rpm. Many less expensive generators run very fast and are very loud as a result. The Onan runs at half the speed and is therefore much easier to be around. When we started using it, we would run an extension cord from the generator to the inverter, which would charge the batteries. This was on the days that it chose to work. Our Onan had a mind and personality of its own, and it inevitably chose to not work on those days that we needed it most. After a week of cloud when we had friends coming up for the weekend, we were just about guaranteed it wouldn't work.

It also had a sick sense of humor. One time it wasn't work properly and I contacted Kevin Moss from Moss' Garage in town. Kevin came out and changed the oil and spark plugs, something that in my books shouldn't have made any difference, and it was purring like a kitten when he left. Two days later it was back to not working. It had a tendency to not run consistently, running too fast or too slow. I wanted it to output 120 Volts, but it would either give me 105 Volts or 135 Volts or something the inverter just didn't like. This was some of the time. Other times, it ran like a Swiss watch.

I eventually got the company that had restored it originally to come up to look at it. They worked away on it for a while, then declared that it probably wasn't worth putting much time into, but that if I wanted to get rid of it, they'd do me a favor and purchase it from me. I make no claim in these matters for being anything but a citidiot, (an amalgam of "city idiot"), but it seemed to me if it was beyond hope, they wouldn't be in such a rush to buy it. In fact, this inspired me even more to pursue getting it working.

Eventually I heard of a local electrician, Jason Elliot, who was experienced in these areas. When in the military, Jason had been responsible for remote power systems. If they were on exercises, and needed power to a field headquarters, Jason had the task of providing it, whether it came from a generator or renewable energy systems.

Jason arrived and within minutes he had the Onan working like a charm. Then he started to show me why non-technical people need to find the right support people in these instances. The first thing he pointed out was that the extension cord we were using was grossly undersized. Once he had rewired it permanently, with heavier gauge wire, we went from getting a charge of 30 AMPS to 120 AMPS coming from the Onan to the batteries. Undersizing the wire was limiting the ability of electricity to flow, so we were wasting generator run time and gas and getting about 25% of it's potential.

Next it was outside to the solar panels. He immediately pointed out the wire that was used for the solar panels. It was undersized again, and was limiting the charge we were getting from the panels. I cannot emphasize enough the importance of finding the right resources, whether this be an electrician or information source. When we moved off the grid there were limited printed materials available. There was a very outdated book on solar power, and a wind power book written by Paul Gipe, but nothing that tied everything together, and took the homeowner through step-by-step, each component of the system, from solar to wind to batteries to inverters to generators, and everything in between. This is why we're so thrilled to have been able to publish William Kemp's book *The Renewable Energy Handbook.*" If we had had this book when we moved off the grid, our lives would have been infinitely easier.

Also, if we'd found Jason Elliot right off the bat, things would have been much easier. I also discussed with Jason the tendency the Onan had to stop working after a while when it was close to zero degrees outside. In fact, there were areas of the generator where I could actually touch and there was frost, even though it was close to the engine. Jason explained this as being "carburetor freezing or icing." It occurs when it's a couple of degrees above or below the freezing point, and moisture freezes in the intake of the carburetor, and it's what was causing the generator to lose power after running for a while. His quick and simple solution was to put a piece of metal ductwork pipe from where warm air left the engine (not the exhaust) and redirect it to where it fed the carburetor, thus eliminating the icing.

The Onan worked well for a year or two, but then fell back into its ornery ways. Jason moved and was unavailable, and we were back to being "challenged" by generators. We had purchased a 6000-Watt propane generator from a friend, that seemed well made, and ran well some of the time, but was very problematic. It didn't like to start when it was too cold outside, so we often moved it into the guesthouse by the fire to get it going. It liked to work well when the propane tank was full, but once it got below about half, it coughed and sputtered and basically said "that's enough work for me." Since we were using barbeque-sized propane tanks for convenience, this meant extra work.

Our neighbor Ken had a 5000-Watt portable gas generator we often borrowed when the Onan wasn't working, but it was tough to get started with the pull cord. Once it was going, the 5,000 Watts really wasn't enough to do the job. It would run most of the loads, meaning that it would keep the house powered, but it wouldn't really charge the batteries very well. When you look at the sine wave that the inverter is putting into the batteries from the generator, the batteries just use the very top of the wave, and if the generator doesn't have enough power, there is very little there for the batteries to make use of.

I eventually ended up with a 6,000 Watt diesel generator, but it had its own set of issues. It was hard to start and wasn't always consistent with the power it put out, so the inverters didn't always like it, and sometimes would just choose to ignore it.

When we upgraded to a new Xantrex 2524 SW, or "sine wave" inverter, our generator challenges continued. The newer inverter was even more demanding in the quality of power it wanted from the generator, which meant our older, curmudgeons of a generator didn't work well. I learned to work around this by putting a portable baseboard heater between the generator and the inverter. This meant that as soon as I switched the circuit breaker to send electricity to the inverter, it first hit the heater. This resistive load was fairly significant and forced the generator to get serious and stabilize itself. The inverter takes several minutes to check to make sure any input source is stable and reliable before it will switch over, and the heater has helped smooth the ONAN so the inverter is more willing to sync up to it.

So we continue to dream of the day we win the lottery. Of the many things we'd upgrade, the generator would be the first. We would get a well-made 10,000-Watt diesel model, which would allow us to run biodiesel in it. We'd also install a combined heat and power system on it, so that we could capture some of the heat that it created, and use it to heat the guesthouse. For now though, we continue to dream. There is one thing that I constantly recommend to people who are contemplating moving off the grid, and that's to not scrimp on your generator. Make it 10,000 Watts, buy a good quality one, and maintain it well.

The other thing I recommend is that you invest in lots of renewable energy inputs to minimize how much you run it. A number of years ago we had a fall that was extremely wet here, which was great to fill up the ponds for skating once they froze, but it meant that we had many cloudy days. We did not have a wind generator that worked at the time so it seemed for a couple of months we were running the generator every three days. Several falls ago set a record for lack of rain, which meant more sun, but as we approach the winter solstice, the days are shorter, there is less sun, and the sun is lower on the horizon and much less strong. We have installed more solar panels and the new Bergey wind turbine, which means very little generator run time. It was quite amazing actually. When the sun came out we were getting 1,300 or 1,400 watts coming in to charge the batteries, and even though we're in the woods, much of the fall was very windy, which meant a steady flow of electricity to charge the batteries.

We now have 2,300 Watts of PV power and we're quite consistently running the generator only two or three times a year. Now during November and December we use our propane stove more than we'd like, but we're burning fossil fuels one way or another. And since more than half of North American electricity is produced from coal, I believe I am having a much smaller footprint from being vigilant about energy efficiency and minimizing my energy requirements, so that I'm comfortable running a generator a few times a year.

With the cost of PV dropping so dramatically since we moved here, for someone moving off grid it doesn't make any sense today not to really bulk up to minimize your generator run time. You know those cloudy days where the cloud sometimes seems thick, and yet you sort of squint, because it's still fairly bright. I'm always amazed now to go out and see our solar panels outputting 500 or 1,000 Watts on those days. Remember they are "Photo" voltaic panels, photo meaning light sensitive. So while they work best in full sunlight, they still will produce some electricity on those cloudy days when lots of light is sneaking through those clouds. Our investment in more PV than we might need has really paid off by dramatically reducing how reliant we are on a generator.

More renewable energy inputs means less fossil fuels burned in the generator, less expense, more independence, and a smaller carbon footprint. Go Solar!

The scary beast that sits in the garage waiting for cloudy weather.

12 The Batteries

by Cam

Over the years we've met people with some unique ideas about what living off the grid means. There is often the perception of dirt floors, outhouses, no running water, and kerosene lamps. The other misperception is that you simply go to bed at sundown when it gets dark. When I presented my fall workshops across the province I often invited participants to come for a short tour of my place so that they could see that I do indeed have lights, TV, internet, and I encouraged them to flush the toilet, just to assure them that we have running water.

Having these incredible luxuries is made possible through batteries, and like all things here at Sunflower Farm we've have our share of stories to tell about our batteries. Some good, some bad. Always challenging, usually humorous.

Most off-grid homes use deep-cycle, lead-acid batteries. These are different from car batteries, which are designed for "cold cranking amps", or a momentary burst of energy to turn over your engine and get it running. Then the battery rests and is recharged by the alternator. Deep-cycle batteries on the other hand get a slow long charge during the day as the sun shines, then have a slow discharge as the stored energy is taken out during the evening, or on cloudy days.

When we bought our place it came with a very large set of ni-cad or nickel cadmium batteries. Most manufactured products are offered at three levels of quality; household, industrial and then military, with military being the gold standard, top of the line level. Well, these batteries were military quality. They had in fact been taken out of a government bunker, one built during the cold war in case of nuclear attack, where the government would have been sequestered in an underground fortification, powered by batteries. Jean, the previous owner of this house, was able to get 60 of these ni-cad batteries when another local off-gridder was getting them. Each cell was worth about $700 new, so we actually had $42,000 worth of batteries. Of course they weren't new, but it still represented a pretty good find.

The beauty of ni-cad batteries is that unlike lead-acid batteries you can't overcharge them, or undercharge them, and they won't freeze. So we were pretty excited about them. And it was a big bank, so it looked like we'd have a good storehouse to get us through cloudy periods. While the batteries were old, they were expected to have a 70-year life. We assumed they went into the bunker sometime in the late 50's, so we hoped to get another decade or two out of them.

There were 5 strings of batteries in the battery room in the guest-house. Each battery was 1.2 Volts, so when you wired 10 together in series you got a 12 Volt system, then the 5 strings were wired together in parallel to increase the storage capacity. Jean had a 6th row in the horse barn where she had a panel charging them, to have lights for when she did horse work at night.

Since we didn't have horses I thought I'd incorporate these into our main battery bank. So with some help I borrowed a trailer and dragged them from the barn into the garage. Three batteries were housed together in one mahogany wood case, which made them basically impossible for one person to move, and just at the limit for two people. Once in the garage I cleaned them up and put a charger on them. Then one thing lead to another and I was swamped with other projects, and the existing battery bank seemed to be working all right, and the extra batteries were just getting in the way. So back to the barn they went.

The original ni-cad battery bank.

Many years later when I put up the Bergey XL1 wind turbine I had to reconfigure the batteries because the Bergey only came in a 24V model, but my batteries were 12 V. I could have done this by removing one of the five strings, and wiring two 12V strings together in parallel, then these two strings together in series to increase the voltage to 24V. Of course, why take the easy route? So the incredible non-portable but constantly migrating and moving barn batteries made their way back to the battery room, with great physical effort as usual. This time I built a very strong wooden shelf to get this string of batteries above the rest, because there was no more room on the floor. Then I was able to rewire three 12V strings together in parallel, then the two 12V strings in series to make 24V but with increased storage capacity.

Of course my inverter was also 12V, which meant I had to upgrade that as well. Oh, and the solar panels were wired as 12V as well, so they had to be redone. On rewire day I got my friend Jerry Horak to come up and my neighbor Ken to come over to help. Jerry went to work pulling the old inverter off, and Ken started rewiring the panels to come in at 48V. I had also purchased an Outback MX 60 MPPT charge controller, and it recommended we bring the DC in from the panels at 48V and then it would drop it to 24V to go into the batteries. The higher the voltage with DC the less line losses you have, so we went this route.

This was one of those days it pays to have friends with technical abilities. When we first disconnected everything I rewired the batteries to 24V, which I had scoped out ahead of time and had all the wiring and bits and pieces ready to go. Then I pretty much did the "tool fetching" routine for the rest of the day. And it's only fair since it's my garage and I know where everything is. Ken also got me to do some rewiring on the solar panels, and in Ken's Socratic teaching way insisted that I understand how I was rewiring them to 48V so it made sense to me. On these days I'm usually in the "can't we just get the darn thing done" mode, but Ken always makes sure I know how to do this stuff myself in case he's not around the next time.

At the end of the day the switch was thrown and everything fired up and worked great. These are always red letter days at Sunflower Farm. There is nothing more gratifying than scoping out a job, getting all the bits and pieces together, and after a day of rewiring and reconfiguring, having everything work properly.

I was pretty convinced after this that the batteries were performing better than ever. This was possible, since we had also added more PV recently, and it was coming in at higher voltage, and we now had an additional string of batteries to increase the total storage capacity. Whether this was me being delusional or they were really working better I'm not sure. But over the next several years it was apparent things were working well.

Bill Kemp had been warning me for a few years that my batteries were at the end of their life and needed to be replaced. I, of course, am cheap and wanted to squeeze every day of life out of them that I could. I was also aware that because they contain cadmium, they are hazardous waste and would need to be disposed of properly.

Then, as so often is the case, fate intervened and helped move things along for me. I was writing our gardening book and needed to take some photos of root cellars, so I put a small ad in our local newspaper. I got a call from someone who lived a fair distance away, but he said that he had just built a new root cellar. We had a long conversation and I realized that he also lives off grid. As we continued to talk we got on to the subject of batteries and it turned out that he had some of the same ni-cads that I had, from the same batch. And he loved them and worked hard to get them. When I mentioned I felt mine were not working as well as I'd like he said 'just say the word' and he'd be up to grab my old ones. It always amazes me when things like this happen, but it was a real stroke of luck.

So I started to do my research on new batteries. I decided to buy them from a Canadian manufacturer called Surrette. They have a number of different models and I really wanted some of their larger batteries. The larger ones had more storage capacity, but with an accompanying higher price. My neighbor Sandy was also in the market for batteries for his off-grid home, so we decided to go together and make the order larger. We each went with 12 2-Volt batteries to give us a 24V battery bank.

The batteries arrived in the winter so we put them on pallets in the guesthouse so I could keep them warm. With lead acid batteries you have to be careful not to let them get too cold. Once I was able to coordinate getting the old batteries removed it was just a matter of having everything in place for that day. On the big day Hans Honegger and Jake from Tamworth came to help, along with Sandy and two people who came to get the old batteries. We had quite a crew.

I shut the inverter and system down and then went about disconnecting the old batteries. They were whisked outside on to a trailer when it was done, off they went. Then we brought the new batteries in and started wiring them together. Then it was just a matter of restarting the system and we were back up and running. The fridge had only been off for about two hours. It was quite remarkable actually.

During the time that I was researching new batteries my ni-cads really started to drop off a cliff, so we were running the generator much more than we had in previous years. So once the new batteries were in it was great to have a good bank again.

As we were moving the new batteries in I noticed some electrolyte in various places. I assumed it had just spilled out of the top as we moved them into place. Then I decided to try and equalize the batteries. This is one area that requires some attention on the part of an off-grid home-owner. While solar panels will just sit there for decades working away, you have to pay attention to your batteries. Over time the voltage in the batteries will start to vary from cell to cell, and ultimately you want to get them all to the same level. One bad or low battery will drag the rest of the bank down. To correct this you "equalize" your batteries, which is a controlled overcharge. You basically set your charge controller to just damn the torpedoes and dump in all the juice you can to really boil any crap off the plates and get those batteries excited!

When I started doing this I noticed an acid smell, which is standard since lead-acid batteries use sulphuric acid as their electrolyte. But then I noticed some liquid on the wooden pallet I set the batteries on to keep them off the floor. The liquid was indeed battery electrolyte as I was becoming intimately familiar with it, after it wrecked all the clothes and gloves I had on during battery moving day. But where was it coming from?

As I examined the batteries more closely I could see it was leaking out of the seam where the plastic case met the top section of the battery. This was not good. Ever get that sinking feeling in your stomach that what you're looking at is going to cause you months of consternation? I get it all the time and I did that day, big time. And it was prophetic. A call to Surrette assured me that this shouldn't be happening, but apparently was anyway. After weeks of back and forth it was determined that a new production line has just been commissioned and the plastic heat sealer was not making an airtight sealed seam. So they would replace the batteries.

After a number of weeks went by another truck arrived and backed down the driveway. I of course had to have the pick up truck ready to ferry Sandy's half down to his house. But as we started to unload the batteries we could see that the plastic wrap that held the batteries to the pallet was covered in electrolyte. Yup, you guessed it, they shipped me the same type of battery, from the same lot, made on the same defective heat-sealing machine.

Now remember that I live off grid for environmental reasons. So the thought of this company shipping me two sets of bad batteries that shouldn't have ever left their factory was a little bit infuriating. More calls to the manufacturer and more excuses, but this time sensing my building frustration my contact suggested that they would ship me the larger, heavier duty batteries, which actually have a double plastic container so they can't leak. This made me happy. Then an hour later he called me back and said his sales manager wouldn't allow him to honor his offer because those batteries were $1,400 more and I'd have to pay the upcharge. At which point all this time of living in the woods and trying to become very Zen-like went out the window and I exploded.

Really, it's not bad enough that they shipped me defective batteries, twice, but then they offer to upgrade me, and then renege? Did they realize that I published the best selling book about renewable energy in North America and I have photos of their product in that book and now they've pissed me off enough that I want to launch a website warning others about these batteries! Michelle came in to the room as I was screaming on the phone and she said "Hang up. Don't waste any more time talking to an employee. Talk to the President." So when I calmed

The new Surrette deep-cycle lead acid battery bank.

down I called the President/owner of the company, who, of course, was out of the country.

The good thing was that I had their batteries. In fact, my garage now had 50 of them, 4 pallets full of defective batteries. Eventually though I got to the President and he upgraded the batteries at no charge. So I ended up with the larger, heavier duty batteries with thicker, heavier duty plates, which I should have ordered anyway. The months of hassle in between are quickly forgotten and the new batteries are working marvelously. We can comfortably go three days with cloud with these new batteries.

These batteries come with a 10 Year Warranty, but if I treat them well they should last 17 to 20 years. And I will treat them properly. I check the electrolyte periodically with a hydrometer to measure their specific gravity, which gives the truest reading of their state of charge. If they hit 50% capacity I have to run the generator. So far with the new batteries I've only had to do that twice, so I think we're back on track to have a minimal amount of generator run time.

With proper working batteries living off grid is a dream. With poorly performing ones it's a pain in the butt. Running a generator is a hassle and I hate doing it. So if you're considering going off grid don't scrimp on the batteries. They are worth the investment.

I so often love to gloat when I hear people complaining about their high electricity bills. I ask them what it's like to have such a bill? I tell them I feel bad that I'm not paying the "Debt Retirement Charge" which they have on their bills to pay off the nuclear reactors built decades ago. I tell them I wish I could help; I really do, but that I can't.

I can't because I just spent $4,000 on batteries! Living off grid, if you want to live a fairly typical North American lifestyle isn't necessarily cheap. If you start now you have the advantage of paying significantly less than we did for your photovoltaic panels. But batteries are still going to be expensive. And while there are all sorts of wonderful lithium-ion batteries in cell phones and laptops and even electric cars, I don't see any breakthrough on the horizon when it comes to off-grid household deep-cycle batteries. You are still going to have to invest some money.

It's money well spent it from my perspective. I have lead-acid batteries which will be easy to recycle when I'm done because the lead can be reused. And I'm not burning coal to make my electricity, so over the life of the batteries I think I'm still having a much smaller footprint on the planet. And you know, it's pretty darn nice sitting inside on a dark winter night, watching a movie on the DVD player, while the fridge keeps our food cold and the pump keeps our water flowing. Batteries are pretty important to make this whole off-grid living possible.

13 Water

by Cam

Water is one of those things that's easy to take for granted, especially if you live in a city. You turn on a tap, out comes the water. You flush the toilet, clean water instantly appears to fill up the bowl. You have a shower, clean water cascades down on you from a fixture on the wall.

It's pretty amazing when you think about it.

In the country, water is a whole other issue. You have to know where your water comes from. You have to go out and find it. Some people have to dig for it, and end up with a "dug well." Dug wells are usually fairly shallow and often have concrete walls. When you think of "wishing wells" from cartoons and literature, those would be dug wells.

Many country dwellers go with a drilled well. This is when you bring in a large drilling machine, mounted on the back of a truck, and you start drilling into the ground until you hit water. If you're lucky, like at our house, after going about 50 feet, you've hit lots of water so you can stop. Some people aren't so lucky and have to go hundreds of feet down, often through rock and at great expense.

Most drilled well holes are about 8 inches across. Once you've drilled deep enough, you generally will put in a casing, which is a piece of perforated material that you slide down the well. This prevents dirt and materials from falling into the well and plugging it up, while still allowing the ground water to seep back into the well after you remove some, for a shower for instance or to flush the toilet.

The key to a good well is the "recharge" or "replenish" rate, or how quickly water will pour back into the well after it's been drawn down, say when you have a bath. As you drill down into the earth you hit seams of water, in the rock and sand and soil. Sometimes you get lucky and have lots of water in your area and sometimes you don't.

One thing that looks likely in the future, is that for many people, getting water is going to take on increasing importance. Where we live in Ontario lakes and ponds surround us, and the province itself borders the greatest source of fresh water in the world, the Great Lakes.

There are many areas of North America that aren't so fortunate. Many southern states have always faced water challenges and now climate change is exacerbating it. States like California and Nevada have huge problems finding enough water for their populations. The Colorado" River which flows through California has so much water drawn from it, much for irrigating crops, that it no longer makes it to the sea and peters out in the desert in Mexico. Water shortages in southern states are going to affect the price and availability of fruits and vegetables for many northerners during the months when we can't grow our own.

Many states in the south western U.S. draw on the Olagala aquifer for their water. This is a huge underground reservoir that provides millions of Americans with their drinking and irrigation water. The problem is that it is being drawn down much faster than Mother Nature is replenishing it.

Atlanta, Georgia has a severe water crisis as drought has left water levels in its reservoir extremely low. Who would have ever thought it would become trendy in urban areas to use your bath water to flush your toilet, but in cities like Atlanta, this is part of the new urban chic.

Climate change is causing precipitation to be inconsistent and unpredictable. While one area is getting too much rain, nearby is in drought. Australia's breadbasket where huge amounts of grain had been grown for the world markets is in its 10th year of drought. Yields have plummeted and farmers are giving up the land.

So regardless of where you live, there's a good chance that water is going to become an issue in your life. Even if you are living in a city, you probably will find water will cost much more. This is for two reasons. The first is that cities are starting to recognize how expensive it is to build water mains and get the water to your home. In many cities, much of their infrastructure was built 100 years ago, and they now have to rip up roads to replace those old pipes. The second is the cost of energy. Moving water around requires huge amounts of energy, so as energy costs increase, so will the costs of products associated with it.

When we think of electricity, many of us think about Niagara Falls and the original source of electricity that was "hydro" electricity, which harnessed the movement of water to generate electricity. If you've ever picked up a bucket full of water, or a case of water bottles, you'll recognize that water is really heavy. So when it's moving it has a lot of energy embedded in it, and it makes an excellent source of potential electricity generation.

The flip side of this is that anytime you move water around, because it's so heavy, it requires lots of energy. In the case of the cities, this energy comes in the form of electricity that the city or municipality must purchase to power pumps, to move that water. As the cost of electricity goes up, so goes the cost of water. The City of Toronto, Canada is a good example. It purchases a lot of electricity. It uses it to power buildings, to keep streetlights on, and for transit. The Toronto Transit Commission has an extensive system of underground subways, and these subway cars are all powered with electricity. It also has electric buses and electric streetcars all powered by electricity. And yet, if you add up all these uses of electricity, the City of Toronto still needs more electricity to pump water, than all its other uses combined.

So pumping water is energy intensive, and for many country dwellers, is one of their greatest uses of electricity. Pumping that water from a hundred feet down a well, and then into the pressure tanks in the house, uses lots of electricity. If you live off the electricity grid in the country, pumping water is going to be a huge factor in your life. In our case, pumping water is one of the single biggest uses of electricity in our off-grid home. It certainly is in the summer when our gardens are at their maximum demand for water.

We have a drilled well with a pump that's about 20 feet down. We also have two large pressure tanks that the pump pressurizes. Our well water comes into our house in the cistern. A cistern is a large concrete enclosure that was common in rural homes many decades ago. Rainwater would be diverted from the roof into the cistern. It was often located right below the kitchen, which made it possible to have a hand pump in the kitchen to bring that water out of the cistern. Cisterns are a brilliant idea and it's too bad we don't use them anymore. Cheap energy has allowed us to use pumps to bring water from wells to circumvent this process.

In many homes with a well, there is a small pressure tank. When someone runs water in the home, the pump comes on and pressurizes this tank again. When the motor in a pump comes on, there is a momentary surge in electrical demand. Once the pump overcomes that surge, it will settle down and draw less current. So one of the first things that should be done in an off-grid home is to install a large pressure tank to reduce how often the pump comes on. When just Michelle and I are living here, our pump usually comes on just once a day. This means that we can go a whole day flushing toilets, washing dishes, filling up kettles for tea and coffee and the pump will come on just once.

Since it's such a big energy user, we've had to make sure we use water as efficiently as possible. This starts with basic things like toilets. We have a low flow toilet that uses 6 litres per flush. There are many new toilets on the market that now actually have two choices for flushing, one for pee and the other for when more of a flush is required. In our case, dual-flush toilets weren't readily available when we first moved in to our off-grid home, so we had a sign on the back of the toilet that requested that users not flush for just pee. It sounds gross to some city folk, but hey, it's just pee. And think of how many ga-zillion litres of water large cities have to process and clean before discharging it back into the environment.

Another use of water in our home is for bathing. We tend to take showers and we have a low flow showerhead, which restricts the amount of water that comes out. This makes less water feel like more. These are the antithesis of the new rain shower type showerheads that basically try and duplicate standing under Niagara Falls. Those are wasteful and environmentally irresponsible. In an off-grid house, they are out of the question, except maybe in the summer when you have power to burn.

Sometimes we prefer to take a bath. Filling up a bathtub uses a lot of water, so in the early days we were very cautious about when we had baths. We actually got to the stage when often if we were running the generator in one of the months when we didn't have enough electricity, we'd also have a bath. After dinner, we'd put the generator on and do the things that used the greatest amount of electricity. So we'd do the dishes, the kids would watch television and we'd have our showers or baths. This way, while the generator was charging the batteries, it was also handling our biggest loads.

There would also be times when I'd be ultra paranoid about reducing water use, so if we'd had a bath, we'd leave the water in the tub. First, because it is a beautiful cast iron tub, the tub absorbs the heat and radiates it back over time to the bathroom. With the bathroom on the north side of the house and a long way from the woodstove, this is a nice way to warm up the bathroom in colder months. Then, when we need to flush the toilet we use the bath water. We keep a bucket in the bathroom, scoop out some bathwater, and flush the toilet with it. My attitude is that we've already used the electricity to get the water this far, and it's going to end up in the same septic tank. Why not put it to one more good use, and have it flush the toilet? If you pour a bucket of water down your toilet, it will take whatever is in the bowl along with it.

Michelle showed me this trick years ago, because toilet cleaning was my job. This was voluntary on my part. I decided if I was going to be a feminist, my daughters were going to grow up thinking toilet cleaning was a man's job. Since we're usually the ones messing it up, it's only fair. I pity the men who end up marrying my daughters if they've grown up being coddled by their mothers and insulated from housework.

Michelle showed me that when you pour a bucket of water down the toilet, it drains the bowl and leaves just a couple of inches in the bottom. So you need another half a bucket to fill the toilet bowl back up. This is important. That water keeps the smell from the sewer or septic tank from escaping into the bathroom, so make sure you fill it up.

Flushing the toilet with a bucket of water sounds very backward to many people, but when you compare it to say an outhouse, it seems pretty nice. And if you've ever lived in an apartment building during a power outage, it's an important skill to have. The pressure in the water pipes in a city will generally take water up to the 6th floor of an apartment building. During a power outage, many cities will have backup generators that they use to keep the water pipes pressurized, because otherwise you have sanitary issues. So as long as you live on one of the lowest 6th floors, you should be fine. But what if you live higher than the sixth floor? The apartment building will have its own pumps that it uses to get water up to the higher floors and most apartment buildings will not have a back up generator for this. So if you're higher, you've got problems. I still remember an image on TV from the blackout in 2003 that left 50 million people in the northern U.S. and Ontario and Quebec without electricity. The TV report showed people walking down 20 flights of stairs to get a bucket of water to flush the toilet, and then walking back up the stairs with the bucket. You've got to be in pretty good shape to do that very often.

When we flush our toilet, it flows to our septic tank. The septic tank has two chambers. In the first, solids settle out, and the liquid part flows into the second chamber. A bit more precipitation of solids happens here, then the remaining liquid flows out to our tile bed, where weeping tiles, or clay pipes (now usually plastic) with holes, disperse the liquid over the "tile bed". The pipes are laid on gravel to allow the liquid to flow, and it trickles down through the gravel, then sand, to be cleaned as it goes.

The solid materials in the septic tank should decompose over time. Every couple of years we have a septic tank pumper come and remove the solid materials that haven't broken down. Septic tank companies have great slogans on their trucks like "We're number one in the number two business."

Ultimately, anything we flush down the sink or toilet, is going to end up in the tile bed, and begins it journey down through the earth, where it is cleaned, and it can ultimately end up back in the water table, from which we draw our drinking water. When you think of this cycle, you tend to be pretty cautious about what you flush. We buy environmentally responsible cleaning products, and use Borax to clean our sinks and toilet. As the toilet cleaner, I will tell you that I will not win any awards for the brightest toilet bowl on the continent. We have hard water, which means there are minerals dissolved in the water. This is wonderful for providing us with a healthy product to rejuvenate our bodies, but can be a challenge in places like toilet bowls where scale can build up.

But anytime I think about calling out the Drano or other heavy-duty chemicals, I remember where it may end up. When city folk start making this link, I think we'll find much better environmental stewardship breaking out everywhere. I can take a slight discoloration in my toilet bowl if it means my drinking water is going to be clean. With the number of people in North America who use bottled water because they don't trust tap water, it would appear we all have to be more careful with what goes down the drain.

I am very aware of using our water efficiently. I remember someone at one of my workshops asking about putting an on-demand hot water tank in his upstairs bathroom, because the water had to run for so long before it got there. Last time I checked I think an electric on-demand hot water heater was $800. I said, "Why don't you do what we do? When we're finished with our plastic apple cider jugs in the fall, we put one by the bathroom sink. Then, when we want hot water, we run the hotwater tap into the jug until it gets hot. We use that water to fill the kettle, water plants, any number of things that we need water for. I could see him processing the information and then I got the feeling he wasn't prepared to go that far. Lots of people in water-challenged areas, especially in the Southern U.S. are already into this habit. It just comes naturally to me now.

We have yet to experiment with grey water, but it's on the list. Grey water is waste water from anywhere in the house except the toilet, which is called "black water." You can use grey water for watering gardens, as long as you use mild soaps.

Pumping water from our drilled well uses lots of energy, so I've spent a great deal of time trying to figure out ways to get water without using electricity. Rainwater is the easiest and most accessible source of water and it can be used for a variety purposes. I use it exclusively for watering gardens, but in some drought prone areas with proper treatment it can be used for drinking water.

One of our first tasks at Sunflower Farm was to get rain barrels on all the downspouts, where the rainwater runs off the roof and into the eaves troughs. There were four downspouts on the house, two on the guesthouse and two on the horse barn, so right away we needed 8 rain barrels. We found a great source of them near our old house, and many of the plastic 50 gallon barrels had actually been used to ship food products, like clam juice. Who would want a 50-gallon barrel of clam juice was beyond me, but once it was washed out, it smelled fine.

These used barrels were around $10 each. I put in a $5 tap and had a rain barrel that can retail in a high end gardening store for $100. Now it's not a designer color, doesn't have a screen to keep debris out or to stop mosquitoes from breeding in it, but our house is surrounded by ponds, so we are going to have mosquitoes regardless of how fancy our rain barrels

Two of our raised rain barrels filled from the roof of the horse barn.

are. It does hold 50 gallons of water, and this is what we want them for.

With 8 of them holding rain, after a heavy downpour we have 400 gallons of water on reserve. Michelle will use some of this water for watering her flower gardens, and I'll drain the barrels down into the garden, using a soaker hose to distribute the water. Luckily the people who built this house in 1888 had the incredible foresight to put it up on a small rise, so it is the highest point around. This helps water run away from the house, and helps to move water from the rain barrels down towards the garden.

I help the rain barrels along by putting them up off the ground, to gain even more pressure. Water pressure is called "head". For every 2.3 feet that water falls, you get 1 psi of water pressure. So if the rain barrel was 4.6 feet off the ground, you'd have 2 psi. When we moved into the house and cut down the huge silver maple at the back of the house, I used large round chunks of the tree to get the rain barrels higher. They looked quite "rustic", but as they got older and started to rot, they looked a little haggard and once they got soft in the middle, they had a tendency to collapse when the barrels were full.

When we had vinyl siding put on the house I knew Michelle wouldn't be happy with hunks of wood as rain barrel holders, so I decided to use rock to build rain barrel stands. We had a large amount of rock from the flagstone walkway that used to run between the house and guesthouse. So I learned the fine art of piling rocks on top of each other to build rock cribs. I put the large rocks on the outside and filled up the middle with smaller rocks and gravel. Now even if the rain barrel isn't there and rain pours out of the downspout, the crib stays in place.

The higher the barrels the better, so now I've even started building some cribs higher than the original 2 or 3 feet I started with. The barrel stand by the horse barn is now almost 5 feet high and you get great pressure when you water with it. I have also started adding second rain barrels to double the amount of water stored at each downspout.

During an average summer, the area around Tamworth, where I garden, is prone to droughts. Speaking to friends in western Ontario, it would seem that so many storms cross the Great Lakes, dump the bulk of their moisture in western Ontario, and have little left by the time they get to us. I've spent way too much time on the Environment Canada website using radar to track rainstorms that seem to vaporize just before they hit us. Kingston will get rain, Bancroft, Perth, anywhere but my garden!

So we're "rain water challenged" here and as the world comes to grip with peak oil, and we all watch our food costs skyrocket, growing some of your own vegetables is going to become an economic necessity for many of us.

We have a garden in the foundation of the old barn on our property and I've built a crib out of cedar at the top of the old ramp to the foundation. The water falls almost 10 feet, and it's pretty amazing how much pressure you get as you stand watering.

Getting the water up to those barrels is another story. There is a dug well beside the barn. When I first started using the dug well I used to throw a bucket down the well, and carry it up to the barrels. I used to sing the theme song from the Sorcerer's Apprentice as I did this. My neighbor Ken likes to point out how I like to do things the hard way. "It's cheaper than a health club membership," I proudly pointed out. That was until my 50+ years started to catch up to me. So now I use a solar panel which outputs 12V DC connected to an $80 DC pump, to pump the water out of the well and up to the barrels. I pull the water from about 10 feet down the well, and then move it to the rain barrels that are 100 ft away, and another 30 feet higher.

Our solar pump working in front of the dug well in the garden.

This is the best illustration of renewable energy making life easier that I've ever had. Solar and wind power produce electricity that does everything in my house from toast my bread, to keep my food cold, but displacing the physical labor of carrying buckets of water up to the rain barrels with a solar powered water pump, was like getting an electrical shock… this solar power really works! Just another "eureka" moment at Sunflower Farm.

From the solar pump, once the rain barrels are full, I use the water for watering vegetables directly. Originally I created my own drip irrigation hose, drilling holes in an old garden hose. On my second attempt I put smaller holes at the end closest to the pump where the pressure is greatest, and larger holes towards the end where the pressure drops off.

Then I purchased a commercial drip irrigation system. This takes the water from the solar pump and distributes it to five 20 ft drip irrigation lines. This way I can water 5 rows at a time without having to move the line each time. I just need to come out in the garden every couple of hours to make sure the solar panel is in the sun and doing its job.

The challenge is that when I want to move the system, I now have to move 5 lines that are connected. Since I'm not using a higher-pressure household water system, I'm able to loosely connect the lines to each other for easy dismantling and moving.

The drip irrigation system works with the capillary action of the soil. If you use a sprinkler to water, you get lots of water over a large area, with a minimal amount getting deep into the soil. You also tend to lose a lot of water to evaporation before and after it hits the ground. Since you want to encourage deep root growth to help drought proof your vegetables, sprinklers can be wasteful. A drip irrigation hose will provide a slow, constant amount of water to the soil. Once it hits the soil it will spread out from the initial contact point and water a larger area using this capillary action. Imagine if you took a large fluffy towel, and kept dipping one corner briefly into a bucket of water. Even if it's in the water just for a second before you take it out, if you keep doing this, eventually that towel will absorb a huge amount of water and it will become completely saturated. Soil works the same way.

With a clay soil, the capillary action will tend to take the water out and away from the drip point more horizontally, while a sandy soil like mine will draw the water downwards. The kit that I purchased had the drip holes at 18-inch intervals, and I thought with my sandy soil I'd need them closer together, since the water won't spread out horizontally as much as clay soil would. The compost and manure I've added to my soil over the years helps the water spread out.

I found that the solar pump is a great aid in keeping the garden watered. Since I don't garden full time, it allows me to fill up rain barrels all day. I go out once an hour or more, and move the water coming out from rain barrel to rain barrel. Then in the morning before I head to the office, Michelle and I can use watering cans to drain the rain barrels and put the water on the rows of vegetables that need it the most. Once the rain barrels are full, I use the irrigation system to drip irrigate a section at a time.

While we have yet to need it, we do have a pond beside the house that we may be able to use for garden watering in the future as the garden along with its water requirements grow. We noticed when we moved in that the spring melt sat in an area beside the house. So we hired a backhoe to come in and dredge out the area so that it would hold more water. Now after a winter of heavy snow or spring rains, we have a small pond. I can see it out of my window in the office and it's wonderful to have. Spring peepers start serenading us as soon as it gets warm enough in the spring. We have several wood duck boxes and often get mergansers to nest there in the spring. And it's home to a large variety of frogs, snakes, turtles and often several muskrats. And nicest of all its gives us a place to skate in the winter. I think skating on a pond beside my house, under my wind turbine is one of the absolutely greatest things about living in the country. The pond is free and the cold weather that freezes the ice is free and it's just a wonderful way to embrace winter.

The final thing I'll mention in regards to water is hot water. We have 4 main sources of hot water that vary with season. I have three hot water tanks. The first is from our solar domestic hot water heater (SDHW). This is the first solar panel most homeowners should put on their roof and it's to heat the hot water they use for bathing and washing clothes and dishes. Using the sun's energy for hot water is actually more efficient than using it to create electricity through your photovoltaic panels (PV), which is why if you live on grid it should be the first panel you install.

We installed a system made by "EnerWorks" which uses a flat panel rather than a vacuum tube system. I prefer this system because it allows snow to slide off it, and secondly because it was designed at Queen's University near where I live. I decided to install it on the roof over our back porch. I didn't trust the roof on our house since it's the original from 1888 and I wasn't sure I'd be able to attach it without major problems.

The porch roof didn't have enough of a pitch so I fabricated my own steel frame to mount it to. This was a pretty big deal for me, because when I was doing it, my neighbor Ken wasn't available to help. With his kind of skill he's always in demand. So I went to his garage with my steel, got him to remind me how to work the MIG Welder, and I was on my own. There's something very satisfying about being able to design and weld a steel frame that's going to house a device that's going to take the heat from the sun and put it into my water.

Once I got the frame bolted to the roof, I got my neighbors Sandy and Don to help me lift the panel on to the roof. It's quite a bit heavier than a standard solar panel and I wanted to be extra careful not to break it in the process.

Then I went about installing the heat exchanger. This is on a hot water tank that is in our bedroom. Some people are appalled to hear that we have a solar thermal hotwater tank in our bedroom, but I felt I didn't have a choice. Our basement floods every spring, so I didn't want it down there. And there is no room on the main floor. Surprisingly when the house was built in 1888 not only were they more concerned about just surviving, they hadn't anticipated ever installing a solar hot water heater.

The flat panel collector from the Enerworks Solar Domestic Hot Water Heater I installed on the roof of the back porch.

The day I first pressurized the system was pretty great. After running for several minutes the copper pipes got so hot you could hardly touch them. And this was in February! Now on the coldest, sunniest days in winter those pipes get so hot you really can't touch them. And now that I'm making my own hot water I've insulated all the pipes so I don't waste a single bit of that heat.

I have a second hot water tank, also in my bedroom, which is my diversion or dump load tank. If I lived on the grid and had extra electricity I would be able to sell it to the grid, but the nearest electricity pole is 4 miles from my house, so this isn't an option. And since I refuse to waste a single watt of electricity that my solar panels have generated, I "dump" this extra electricity into my "diversion" load, which is a hot water tank. So the excess electricity goes into the heating elements in the tank and heats water.

My third hot water tank is downstairs, under the other two tanks, in the pantry beside the bathroom and it's propane. Now this tank rarely comes on. It does sometimes in the fall during the dark months or in the spring during those rainy weeks, but most of the time the sun is heating my water. Our goal is to pull the plug on this tank all together. To do that we'll need to tie our woodstove into the system, since the woodstove is usually on during those periods when the propane comes on. This would eliminate our use of propane for hot water completely.

In the meantime we use the woodstove for hot water anyway. We always have a kettle on it for our tea, and we put extra water on if we're having rice or pasta for dinner. And on bath nights, I get the stove loaded up with 3 big corn pots and our biggest kettle. I try and have them on top of the woodstove by noon if we're having a bath that night. I also usually fill up 5 or 6 buckets of water that I carry to the living room and let them warm up by the woodstove. Whether your water comes from a municipal supply, or out of a well like mine, it's very cold when it comes out. By getting it into buckets I let it get closer to room temperature all day and no energy is required for this, other than the wood we're burning in the woodstove. The other advantage of this is that often I'm doing this when the sun is on the panels as opposed to doing it at night and having the energy for the pump drawn out of the batteries.

So by evening when I fill up the tub with the pots on the stove, that water is scalding hot. Then I add the 5 or 6 buckets of cold water that's warmed up to room temperature and the water is still often too hot for Michelle to get in. After Michelle is done 15 or 20 minutes later I use the same water. In this day of hygiene obsession some people just aren't comfortable with this concept, but I've been doing it for years. All this energy has gone into heating this water, why wouldn't we use it for more than one bath? It's just second nature to me. And then the following morning I use that now cold bath water to flush the toilet. I do this because it just makes sense to me since it's going to end up in the septic tank regardless; I'm just rerouting it through the toilet first. I only do this at the time of the year when power is at its minimum, but it's so second nature to me it's hard to actually pull the plug and let the bathwater drain all on its own during the other times.

Our friend Bill Kemp has completely automated all his systems, including running his hot water through his wood cook stove and woodstove, and he describes in it great detail in "The Renewable Energy Handbook." He thinks I'm taking a very backward approach to things like this by doing it manually, but it's working for me. It's good to know that I can strive to automate it in the future. It's a technology I like though. I carry the water from the woodstove. It works.

One thing that I love about living off the grid is that it keeps me in touch with how precious things like water are. It is essential for life. It is all recycled through the eco-system so you become intimately aware of what you're putting down your drain. And when you realize how much energy is required to pump water from the ground into your house, you start to treat it like the amazingly valuable part of life that it is

14 Bungle in the Jungle

by Michelle

One of my favourite aspects of living in this rural area is the opportunity to share my surroundings with a variety of animals, both of the wild and the domesticated type.

Our animal collection began within hours of moving into our new home. After officially moving in that cold January day, our new neighbors, Ken and Alyce, invited us over for dinner. The temperature was in the mid 30s – that's **minus** 30° Celsius (-22°F) – and as our tires crunched on the cold hard-packed snow of their driveway we spotted a kitten running across our path. The girls jumped out of the car and ran to our neighbors' door where the kitten was waiting to be let in. They scooped the kitten up in their arms, oohing and aahing over it, as we were welcomed into a warm and bright kitchen. We admired the little tabby and asked Ken and Alyce about it. It turned out that it wasn't their kitten – it had shown up at their door a few days previously. They told us that someone had been living at the campsite on their lake in a camper and had finally packed

up and left, leaving behind this cute little kitten. They said "She's yours if you want her ..." So, on our first official night in our new home, we returned from dinner with a new member of the family! Poor Thomas – not only had our first cat been moved from the comfort of his city home to this new place in the country but now he found himself having to share his new home! Luckily the two of them hit it off pretty quickly and they became the best of friends. In fact I've never known two cats to "cuddle" as much as they did and greet each other with kisses (bumping each other on their foreheads).

Emma, as she was christened, provided us with lots of new experiences. As it turns out, she wasn't so much a "kitten" as a malnourished cat and once she was provided with a steady healthy diet she quickly matured and went into heat! Suddenly we found ourselves with a yowling, frisky cat. Thank goodness Thomas had been neutered! However Emma took matters into her own paws and ended up disappearing for a few days. We were sure that some predator in the woods had eaten her but luckily she reappeared looking none the worse for wear. The local "fishers" are notorious for snacking on pet cats! Fishers are ferocious weasel-like predators that were introduced to our area many years ago to try and deal with the porcupine population. Fishers are one of the few animals that will kill a porcupine.

We immediately scheduled an appointment with the vet for Emma's spaying. However, after dropping her off at the vet's office on the appointed day, we received a call from them to tell us upon examining her they realized that she was already pregnant and we were given two choices – they could carry on with the operation, or we could come and pick her up and let her have her kittens. We decided on the second alternative.

We watched our tiny little "kitten" expand quickly and we provided her with a cardboard box "maternity bed". One morning at 2 am we were woken by the sounds of Emma in labour. The girls and I got up quickly and tried to encourage her to move into her maternity bed but she had other ideas! She tried to hide under Nicole's bed in her fright, but we were able to convince her to stay on top of the bed where we could help her. In the end she gave birth to five kittens on the top of Nicole's bed (luckily we were able to quickly put down some old towels etc.). We were all pretty excited to have played "midwives" to our cat!

As soon as we moved to our country home, everyone seemed to assume that we'd be acquiring a dog. Cam was raised without any pets at all and so he hadn't had much experience with dogs. I had been around dogs but hadn't ever owned one myself and in fact would consider myself more of a "cat person". We weren't in any big rush to acquire a dog despite the fact that we had plenty of opportunities to. Shortly after taking possession of this house we spent a weekend here. We had some friends with us and at one point everyone except me had gone swimming in a lake down the road. I happened to look out and noticed that a dog had arrived at our house. He was lying on the front porch as if he owned the place. I was reluctant to go outside with a strange dog out there – what if he was vicious? My family and friends arrived home shortly afterwards and were greeted by a friendly wagging tail. After determining that he was not vicious we decided that we'd better feed him – after all, who knew how long he'd been traveling on his own. We didn't have much to offer him – some tofu hotdogs, some buns and a banana. He ate everything except the banana – I guess he wasn't THAT hungry! We called Animal Control and when the worker arrived he informed us that this dog was a purebred German Shorthair Pointer. He seemed surprised that I would have called him and not just kept such a valuable dog for myself!

Morgan the Wonder Dog

This was just the first of many, many dogs that have shown up on our doorstep over the years. Most are hunting dogs that became separated from their hunting owners. Usually a call to the local Humane Society sorts these dogs out.

After we'd been here for a couple of years we got a call from a friend in Burlington. He had a dog that needed a new home and wondered if we'd be interested. We agreed to stop in the next time we were back in Burlington and so a couple of months later we found ourselves meeting "Morgan" a Shetland Sheepdog. Our friend's son who was living in an apartment in Toronto at the time had purchased Morgan. The breeder who sold Morgan told him that he would make a "great apartment dog." Anyone who knows anything about Shelties knows how misleading this statement was!

Shelties have a seemingly infinite amount of energy and Morgan did not take kindly to being cooped up in an apartment all day long. Soon Morgan was living with our friends in their suburban home but even that small yard wasn't enough room for a Sheltie!

We liked Morgan right from the start and after talking it over on our way home we agreed to take him. Our friends dropped him off a few weeks later. Morgan jumped out of their van and began running around our property. He even jumped into our pond! We asked our friend "Does Morgan swim?' and our friend replied, "I have no idea but I guess we'll find out now!"

When it was time for our friends to leave for their return to Burlington, they were worried that Morgan would jump back in the van and want to go home with them. They needn't have worried as Morgan was having too much fun exploring his new, expanded horizons and he never looked back!

Morgan has been with us for 10 years now. He is a wonderful dog and does everything you'd like a country dog to do – barks to warn us when anyone arrives, keeps the wildlife out of the gardens and inspires us to go for walks! In the spring, summer and fall he voluntarily sleeps on the front porch. Once it gets cold enough in the fall he chooses to come in to sleep inside at night. But he is mostly an outside dog and prefers to spend most of his time outside.

He also relishes responsibility. Every year we expand our gardens and we grow more and more corn, with the hope of selling some of it. Somehow, right before we're ready to harvest it, a gang of raccoons will get into it and do a real number on it. It wouldn't be so bad if they'd just knock down a stalk and eat that ear, but they like to knock down whole sections and take one bite from each ear. So now as we get close to harvest time Morgan "The Wonder Dog" sleeps in the garden. Cam takes him out at dusk and explains the importance of his task. He is rewarded with a piece of bread, which is one of his favourite treats. With Morgan on duty in the corn patch the raccoons don't come anywhere near the corn. At sunrise we will hear Morgan barking, letting us know that his shift is over. We head out to the garden to let him out and we give him another piece of bread and praise him for his work. I believe this is one of the reasons that dogs were domesticated. We love Morgan and he is a very important part of our family, and he is an integral part of our food production as well!

One October morning I was returning from an early morning trip to town. I was about to turn into my driveway when I happened to notice two very cute little faces peering at me from the side of the road. I tried to figure out what I was looking at – were they raccoons, foxes, or what? I parked my car at the side of the road and got out and started to walk towards these creatures. Suddenly I realized that they were puppies and

in their excitement to see me they almost jumped into my arms!

I noticed that the mother was hanging back behind them and when I scooped them up to carry them to my house I tried to convince the mother dog to follow us. She would have no part of my plan however and with just a quick glance back at her puppies she ran off. My concern was with the puppies and I hoped that she would eventually choose to join them. I found a big cardboard box with tall sides and put a dish of water and a dish of dog chow into it. The puppies devoured the food and the water in no time at all and then curled up to have a sleep.

As it turned out, the puppies showed up on the day we were pouring the cement for our front sidewalk and so I wasn't able to spend as much time with them as I would have liked. However, one of our friends who showed up to help with the sidewalk ended up deciding to take one of the puppies and so the timing was serendipitous in that respect. The other puppy was adopted by one of my daughter's high school teachers and so that story had a very happy ending.

As I mentioned, we've had many, many other dogs show up here. Many are tired and weak from having run many miles looking for a human after they've become separated from their hunter-owners. In all the time we've lived here we've only had one stray cat show up. It was a short time after moving in and I started to catch glimpses of a white cat that would run through the yard but never come near. I jokingly called it "Phantom" and suggested that it was a ghost cat. I'm usually pretty good at winning over stray animals but this cat would have no part of my efforts to befriend it. It became problematic when it started picking fights with Thomas and Emma whenever they were outside and so I really hoped that we would be able to find a new home for it. My daughter Katie eventually had the magic touch that brought this cat around and we were able to catch it and take it to a new home in a village south of here.

For our first couple of years here our cat population consisted of Thomas and Emma. They were mostly inside cats that went out only occasionally. When Cam began to notice an ever-growing mouse population in the barn and the woodshed, we decided to adopt two ginger tabby kittens. We hoped that they would live in the barn and help to keep it mouse-free. They came from a farmer south of here who often has excess kittens that he is willing to part with. These two were almost wild and very feisty. Ginger, as my daughters named her, was longhaired and her sibling, whom the farmer assured us was a male, was shorthaired. My daughters named him "Rusty". We made the kittens a home in the barn and began leaving kitten chow out there for them. Very soon I became amazed at how quickly these two little kittens seemed to go through food. Then I noticed that their dishes were often muddy and this surprised me because I had never known kittens or cats to put mud in their food dishes. Finally one day when I went out to feed them, a raccoon happened to be leaving the barn as I was arriving. Ah ha! I realized that the local raccoons had discovered a cheap meal in the barn.

As the kittens grew up, they managed to protect their turf without my help. Ginger was a feisty little cat and one night at dusk we heard some cat screams and growls coming from outside. We ran outside to discover Ginger and a raccoon tangled up together like a big fur ball, rolling across the lawn. Apparently Ginger got the better of that raccoon as it raced off as soon as it became untangled from her grip!

At about this time I started to question the gender of little Rusty. He wasn't looking much like a "he" and sure enough, I realized that our farmer friend had led us astray. "Rusty Boy" became "Rusty Girl" at this point! As the kittens approached sexual maturity I took them to a vet to get spayed. As much as we love kittens, I didn't want to have an ongoing supply of them!

Unfortunately, both Ginger and Rusty eventually disappeared and we can only assume that some local wildlife got hold of them!

At this point in time we share our home with 3 cats. Thomas has passed on and Emma is the "old lady" of the bunch. Cricket is a black

& white cat that joined our family as a kitten when one of my daughters received her as a Christmas gift one year.

The youngest of our cats joined our family in a very unusual way. One cold and miserable February afternoon I had just pulled out of the grocery store parking lot in town and as I drove along the wet and slushy roadway I noticed something furry running down the middle of the road. There were 4-foot snow banks along the side of the road at that time of year and I immediately worried about how the little creature would manage to get off the road if a vehicle came by. Luckily there were no other vehicles on the road at that moment so I stopped my car and began to step out of it. The furry little creature, which I quickly recognized as a kitten, jumped in to my car, up on to my lap and then up on to my chest where she proceeded to cover my face in little kitty kisses!

Despite my efforts to find the owners of this wonderful little kitten, I wasn't immediately able to figure out where she had come from and so she came home with me. She was eventually welcomed by the others and has become a very treasured part of our cat family! The mystery of her origins was solved one day when a local acquaintance dropped in. She saw Lizzie and asked, "Where did you get this kitten?" I began to describe our rather unusual meeting and she said, "I think this used to be MY kitten!" I gulped and expected her to demand that I return her kitten to her but luckily she went on to share the story of Lizzie's beginnings. It turned out that Lizzie was born on a farm about 10 kms (6-1/2 miles) north of town. Her mother disappeared leaving behind a litter of kittens that were much too young to be weaned. My friend ended up adopting Lizzie and bottle-feeding her until she was old enough to be weaned. At that time she took Lizzie back to the farm and hoped she would be able to survive but Lizzie disappeared within the week! Both my friend and the farmer's wife worried that Lizzie had met the same end as her mother had. Apparently not. Somehow Lizzie managed to travel the 10 kms in to town to the spot on the roadway where I met up with her.

As soon as we moved in to this place we started noticing how empty the paddock looked. Our neighbor Alyce didn't have her own barn at the time and was boarding her horse at a friend's place. She let us have her horse "Toby" stay here from time to time and we found that we enjoyed having a horse around.

Alyce was talking to our neighbor Agnes, who is a beef farmer down the road. She had a horse and a donkey living on her farm and she was willing to lend them to us with the condition that if we ever wanted to give them up, we would give them back to her. So Abby the horse and Dorothy the donkey came to live here on Sunflower Farm. They had been living in a paddock full of burrs so the first priority was to give them both a good brushing and clean up to get rid of the burrs.

Abby suffered from COPD (Chronic Obstructive Pulmonary Disorder) and had quite a cough. Agnes had sent along some cough syrup for her and I mixed this into her sweet feed once or twice a day and it seemed to help.

We enjoyed watching the two of them graze in the paddock and we let them come out and mow down the lawn whenever we were outside and could keep our eyes on them.

A few months after arriving here, Abby passed away. She had seemed fine in the morning and ate her usual meal of hay and sweet feed. At some point during the day I noticed that she wasn't out grazing in the paddock and I went out to discover her lying in the barn. She didn't seem to be in pain and appeared to just be very tired and in need of a rest. As the day went on, I took buckets of water for her to drink and continued to keep an eye on her. Alyce arrived here later that evening and together we watched as her breathing became slower and shallower and as we stroked her and spoke to her she eventually stopped breathing.

I felt sorry for Dorothy who had lost her companion. Alyce allowed us to have her pony "Tique" here to keep Dorothy company. Tique was

a spirited little pony that seemed to think that the grass on the far side of the fence was greener and tastier. One evening we had just sat down to have some dinner when we heard a CRASH! I looked outside and noticed Tique out grazing on the lawn. I couldn't imagine how he had managed to escape until I saw the barn door. It wasn't the big animal-sized barn door but the smaller, wood and glass door that allowed people to go in and out of the barn. Tique had obviously been admiring the green lawn through this glass and decided to leap through the glass in order to take advantage of the nice green lawn.

I called Alyce and she came over to assess what damage Tique had done by jumping through a glass window. Luckily he had only a small gash along his side that she was able to clean. Ken helped us to repair the door with a spare one from his own place and we returned Tique to the barn and locked the door behind us!

Eventually Tique left us and Dorothy was alone again. Then Alyce caught wind of a horse that needed a new home. Destiny came to us from a home where she'd been neglected. She was skin and bones when she arrived but she soon filled out on a high fiber/high fat diet from the local feed mill. Once again Dorothy had the company of a horse and we enjoyed watching Destiny regain her health.

Since none of us are riders, Destiny and Dorothy were just big "pets" or "hay burners" as Cam liked to call them. They were a lot of work requiring feeding both morning and night as well as the chore of cleaning up after them in the barn and the paddock. Cam was grateful for the constant supply of manure for our gardens but the truth was they were a lot of work for a small amount of fertilizer! Since we didn't grow our own hay we were faced with sourcing and buying hay a couple of times a year which included going to pick the hay up, loading it into our truck and then unloading it into our barn. This was a great workout but a bit inconvenient.

There was also the matter of a farrier who needed to come by 4 times a year or so to trim Destiny's and Dorothy's hooves. Without trimming, their hooves grew uncomfortably long and unhealthy. There are a number of farriers in this area, but we found that none of them were too anxious to drive this distance to trim just two animals' hooves. So I spent too much time on the phone begging and cajoling a farrier to come and this became just one more annoying aspect of having large pets!

As Nicole and Katie headed off to university and Cam and I became busy traveling on weekends to promote our books we realized that it was time to find new homes for Dorothy and Destiny. Once again Alyce came to our rescue and helped us to transport Dorothy back to the beef farm and she also found a new home for Destiny.

We've hosted other horses and ponies over the years as our neighbor Alyce juggles finding homes for various rescued animals. Now that she has acquired a herd of Highland cattle, we have hosted one or two of her cows from time to time. When Billy the bull was too young to join his new herd he spent some time in our paddock getting bigger and older. A female named Aggie was here keeping him company. Aggie's mother rejected her at birth and so my neighbors ended up bottle-feeding her. She thinks she is a pet and is very friendly and affectionate. Once Billy was old enough to join the herd and hold his own, our paddock became empty once again.

Our latest additions to the menagerie are four Rhode Island Red chickens. Henrietta (Henny), Penelope (Penny), Flora and Belle have taken up residence in a coop that Cam built which is temporarily on our front lawn. We still like to eat eggs and we can't always find a local farmer who has eggs we can purchase conveniently. So we decided that if we are going to eat eggs we'd better know that the chickens are raised humanely.

Commercial pressure for cheap food has resulted in most chickens being raised in cages of 6 to 8 chickens, each chicken occupying as much space as a sheet of 8-1/2" x 11" paper.

Cam originally ordered two 20-week-old chickens. This is approximately the age they start laying. Once we got a feel for how much work they were going to be we decided that it made sense to get two more chickens.

We have really enjoyed getting to know our chickens. They rush out of the coop at sunrise when we open their locked door. They spend the day pecking and scratching the earth as chickens are meant to, looking for bugs and other things to eat. Periodically we let them out to roam the property when we have time to supervise them. At dusk each night they put themselves to bed. They make their way into the coop and perch on their roost.

We are hoping that they will provide us with about two dozen fresh eggs every week. We hope to be able to start selling some periodically as well.

Of course we've seen our share of wildlife while living here. Sharing these woods with wild creatures is a highlight of living here for me. In the years we've been here we've seen bears, foxes, a wolf, coyotes, deer, a

moose, fishers, porcupines, skunks, raccoons, flying squirrels, chipmunks, mice, groundhogs, muskrats, beavers, otters and lots of birds including turkeys, bluebirds, owls, ospreys, hawks, and hundreds of songbirds.

Some of these animals are plentiful and you see them quite often. We often see deer on the road and occasionally they will come close enough to the house without Morgan noticing that we have watched them out the window as they grazed on the lawn or nibbled the leftovers in the vegetable garden. Luckily I've never had a problem with them eating my flowers but I know other country dwellers do have to deal with this.

We also see quite a few red foxes and in fact for 2 years in a row a mother fox has used a small cave in the side of the road right across from our place as a den. We would often see her kits scrambling back into the den when our car approached and once Katie and I were able to go out and actually play with the little pups while their mother was away. It was just like playing with puppies, as they would chase a stick if I dragged it across the gravel and they were quite curious about us and came quite close to us to investigate.

We don't often see coyotes but we know that they are around. Morgan will often bark when he is outside at night and when we go out to investigate we'll hear the cries of many coyotes as they howl and yip at each other.

Cam spotted a wolf early one June morning as he headed down the road for a walk. As he walked along he watched it bounding through the brush at the side of the road and then it landed on the road right ahead of him. As soon as it caught sight of Cam it was off, bounding back in the direction it had come from. Interestingly enough, Cam's encounter with this wolf happened right around the time of his mother's death, as he discovered later when his Dad called. An interesting coincidence?

The moose wandered into our yard early one spring morning. It was about 5 am and we were sound asleep when Morgan started to bark. It was a different bark than we'd heard before and so Cam got up to investigate. He didn't have his glasses on as he looked out our bedroom window but even without glasses he could tell that there was "something big" on the front lawn. I grabbed my glasses and looked out and there stood a moose about 30 feet from the house. Morgan was giving it a fair amount of distance but was barking and barking at it. The moose looked down at Morgan with a look of disdain before it finally plodded through the brush and across the road. Cam tried to grab the camera but it was gone before he could get a picture. We went exploring on the land across the road and found the big hoof marks that the moose had left behind in the mud.

When we first moved here the raccoon population seemed enormous and they were not the least bit intimidated by us. I was once on my knees gardening and felt the presence of something behind me. I looked behind and there sat a huge raccoon about 3 feet away. It seemed intrigued by me and was content to sit and watch me. Unfortunately I found the experience a little unnerving.

As I mentioned earlier, the raccoons started getting into the cat food and fighting with the cats and so we decided it was time to move them. We began using a live trap, baited with Oreo cookies smeared with peanut butter, and once we trapped a raccoon Cam would load the trap into the back of the pickup truck and head down the road to a less inhabited (at least by humans) area. One night he trapped a rather large and very aggressive raccoon. Cam decided to throw a blanket over the cage when dealing with this raccoon – it was that aggressive! He loaded the trap into the back of the truck with the blanket still on it and headed down the road. He met Ken and Alyce along the way who were out for a walk. Ken was walking a dog and Alyce was riding a horse. He stopped and Ken asked

what he had in the back. Cam began to describe this enormous, ferocious raccoon that he had trapped and Ken started to lift off the blanket to take a look. Cam warned him and suggested that he might not want to tangle with this raccoon. When Ken removed the blanket he began to laugh – the cage was empty!! The raccoon must have escaped and jumped out of the truck when Cam wasn't looking. This has become one of the many stories that Ken likes to tell his friends - the night he met Cam on the road taking his empty raccoon trap for a drive.

Cam also caught a skunk in the raccoon live trap. He made it very clear to me that he wasn't going to try and remove it. So I was left with the responsibility of releasing it. I approached it with great trepidation, but I was slow and careful to not make any quick movements and I managed to open the door and let the skunk go free without incident.

One time Cam decided to move a porcupine that was close to the house. While porcupines eat bark and don't pose any real threat, it was close enough to the house that we were concerned that Morgan might go after it and get a mouthful of quills. The porcupine was perched near the top of a small poplar tree. So Cam donned his thick chainsawing pants, cut the tree partially through and lowered it slowly. He then put a garbage can over the porcupine, then slid the lid underneath it and turned the can over slowly. He loaded the porcupine in to the back of the truck and "Cam's Wildlife Relocation Service" made another trip to the uninhabited bush for another "catch and release!"

Cam was also the first person to experience a flying squirrel. One evening, right at dusk, Cam was out on the driveway and heard the sound of something scratching the metal tower (the one that our phone antenna is on.) He looked up and saw a squirrel and wondered what a squirrel would be doing on a metal tower. He had barely given it any thought when the squirrel leapt off the tower and swooped down towards him. I heard a scream as the squirrel managed to glide past Cam and into the woods. Needless to say Cam was shocked by its behaviour.

When we moved here there was a small depression in the ground near the house. It filled up with water in the spring and we had a pond! Unfortunately as the hot, dry summer passed, this little pond had a tendency to dry up and so we eventually had a backhoe come in and dig the pond out. We now have a year round pond there and have enjoyed the wildlife that has made the pond their home. A family of muskrats built a home at the side of the pond and this spring they were joined by a young beaver. We were surprised to see a beaver, trying to survive in our small pond but it was cute to think of this beaver living with muskrats! Sure enough as the summer drought caused the pond to disappear, the beaver disappeared too.

My Dad built two wood duck boxes for beside the pond. Mergansers have used them and raised many clutches of eggs in them. Cam can see the pond from his office. One day he thought the merganser was cleaning out the box because he could see something being tossed out into the pond. It turned out that Mother Merganser was tossing her babies out of the nest in to the pond below. We often saw the mother leading a procession of merganser babies around the pond.

Cam's view of the pond gives him a front row seat for many nature events. One winter day he watched a fox working away near a large pile of brush. Eventually the fox dragged a rabbit out and proceeded to eat it. Foxes are carnivorous predators, and as cruel as it may seem, this is what happens in the wild.

Living in harmony with the many creatures that call these woods their home has been one of the biggest joys for me. Having the space for "big pets" like donkeys and horses is wonderful, and being able to enjoy the sights and sounds of the abundant wildlife is something that I will never tire of.

15 Our 100-Foot Diet

by Cam

Back in 1976 when I was in high school I was into all the usual stuff. My favorite band was Boston, but I liked Kansas and Chicago. There was a geographical theme to band names back then. I loved the freedom that access to my parents' car provided. It helped me to pursue another interest—girls. Part-time jobs were handy for gas money. I tried to do well in school and took all those academic courses to expand my brain. It wasn't until I bought an off-the-electricity-grid house that I realized what a huge mistake I had made by not taking electrical, automotive, and other practical courses that would have actually helped me to learn a few skills.

My interests were girls, music, cars, and school, in about that order. Then I did this weird thing. I looked at my parents' backyard and thought: I'd like to grow a garden. I wasn't into the groovy Woodstock music thing, so this wasn't some hippie back-to-the-land urge. And it wasn't a message I'd received during a chemically induced hallucination either, since I just never could suck that smoke into my lungs. I lived in suburbia. I grew up on cartoons and The Partridge Family. I'd never been to a farm. All my ancestors were city dwellers.

Michelle and me gardening in our younger years.

Nope, this was something beyond that. I have yet to figure it out. As I ease on down the yellow brick road of my spiritual life I think perhaps it was motivated by a former life. Perhaps I was a farmer in a previous life and this was part of an unfinished journey. Maybe some force in another dimension was directing me. But it seemed to be something stronger than that. It seemed to be something at a cellular DNA level. This was something rooted deep within me that I needed to deal with. It was completely out of context. My normal activity at the time was seeing how loud I could play "Stairway To Heaven" by Led Zeppelin on the insanely cool and high-tech cassette deck in my parents' car without blowing up the speakers. What the hell was with the gardening thing?

The gardening thing wasn't to be ignored though, so I took the calling to heart and put in a vegetable garden. I took a shovel and I turned over the sod. The subdivision where I lived had soil that was very high in clay. Clay can be pretty hard stuff to grow in. Back in the 70s when this subdivision was constructed they were still learning, so the developers scraped off all of the topsoil to make it easier to dig foundations and build on a larger scale. They didn't put the topsoil back though. If they had I would have had bad topsoil. Instead what I got was abysmal clay subsoil. The clay content was so high I would have been way better to buy a potter's wheel and start making ceramic bowls.

Luckily at the age of sixteen I was too stupid to know this, so I persevered. I added a bag of peat moss hoping it would help break up the clay, which was like throwing a granule of sugar in to sweeten your coffee. Then I planted some seeds and waited. Things did grow, but they did not grow well. After a rain I'd have to go out with the hoe and break up the soil. The rain turned the surface of the garden into a concrete-like substance, like ice on a frozen pond. What this garden needed was a jackhammer to break it up. But a humble hoe and teenage bravado were all I had, so I kept at it.

The results were abysmal. Very few things grew and what did grow did not grow well. An intelligent teenager would have given up there.

But lo and behold the following year we moved to a different city and this time to a home closer to the downtown. The home was a hundred years old and it turns out that a hundred years ago they didn't have fancy earthmoving equipment and limitless fossil fuel energy, so they left the soil that was there alone. Since many cities were actually established near bodies of water, you'll find city soil can be pretty good. The soil at this house was fantastic. It was dark and rich and wonderful. The only problem was that there was a patio right where I wanted to put the garden. My parents were great. No problem, they said. Just move the patio stones closer to the house and that area is yours. Do you have any idea how heavy patio stones were in the 70s? They used concrete and lots of it. And reinforcing rod, so the darn things were basically indestructible.

Once again the strength of my teenage back was undaunted and the patio stones began their migration. The soil under the stones was perfect. I turned it over and raked out the weeds and in went my seeds. This time it was a huge success. One of the photos of the garden captured my joy at the prospect of growing in real soil. It also captured my Peter Frampton-like hair. All right, it wasn't that long. I notice that I kept my seeds in a North Star shoebox. Back then you wore Adidas or North Star. I believe North Star was the poor cousin and therefore my shoe of choice.

But having a garden where things actually grew was pretty novel. Having my parents rave about how great the tomatoes were was a bonus. Who'd have thought that? You could put in a tomato plant in June and have tomatoes in August. Not that I liked tomatoes. Not that I liked any vegetables for that matter. This was the weird thing about my gardening pursuit. It wasn't as if the outcome of growing vegetables was something that I particularly desired. Now if I could have figured out a way to grow Cap'n Crunch or Cocoa Krispies, then it would have made sense. But I couldn't, and frankly all these healthy green vegetables weren't even on my radar screen for preferred dinner items. Thankfully 35 years later they are indeed the basis of my diet and I enjoy them immensely.

After a year or two away from a garden while in apartments, the bug finally bit me again and this time I discovered a place where I could rent a garden plot in a community garden. The garden required a drive in the car to the other side of town, which was a bummer, but after dinner many nights and on weekends I spent a lot of time there. The garden was actually in Bronte Provincial Park, which was an oasis in a concrete jungle. It was peaceful and quiet and I was surrounded by green. As I worked in the garden I could hear birds. When I'd finished I could walk in the woods. I feel blessed to have had this outlet for my gardening obsession. Some forward-thinking park warden had had the vision to realize just what an asset a publicly owned trust could be and what benefits it could offer its constituents.

While the garden was wonderful it was communal, which did have a few downsides. The biggest was the neighboring plot gardeners' approach to pest control. Some seemed to think the images of Agent Orange spraying in Vietnam were a model to be emulated and felt 2,4-D was a better way to eliminate weeds than a hoe. I still retain the image of a gentleman dusting his potatoes with this thick, white cloud of some unknown toxin that looked like the DDT they covered prisoners with in those grainy black and white movies from World War II. Did he not think that if it could so easily snuff out a potato beetle maybe, just maybe, there might be some long-term effect on him personally? And this was before I had heard about or read Rachel Carson's *Silent Spring* or any of the supporting evidence that was beginning to show a link between these products and negative outcomes in humans from long-term exposure. It just didn't seem right. And it just didn't seem necessary. Oh look, there's a potato bug. How about I squish it? What? No potato bugs? So maybe I won't take precautionary measures and nuke the plants regardless. That was the wisdom of the day: precautionary application of pesticides to avoid potential problems. And we wonder how insects developed broad-spectrum resistance to so many of these chemicals.

I was blessed to finally be in a situation where I could buy my own home, and Michelle and I finally got a place to call our own. As the Dixie Chicks say in "A Home," "Four walls, a roof, a door, some windows, Just a place to run when my working day is through." And it was a small home. I think about 800 square feet on the main floor and a basement that was great for storage. The lot was small, 40 feet wide by 100 feet deep, but it was ours! That first spring I got to put in my first garden on my own property. It was small but it was an older neighborhood so the soil was good. I knew it was going to be a challenge because the wonderful huge tree that shaded our non-air-conditioned house also made a fair amount of shade where the garden was. But I put it in a far back corner and hoped for the best. The best turned out to be pretty dismal and nothing grew well. The soil looked fine. It looked rich and should have had the nutrients the plants needed. I kept it watered and weeded and it got enough sun that there should have been some success, but it was pretty much a disaster.

This is where we learned about research and after some digging realized the culprit was the big tree. And it wasn't even the obvious explanation, which was the shade. The tree was a black walnut and for a vegetable garden that's a death sentence. Black walnut trees, and I mean the whole tree, the leaves, the roots, the nut husks, secrete something called juglone. Juglone is a "respiratory inhibitor," and who'd have thunk it, plants need to respire just as you and I do. So while my tenacious tomato plants were trying to grow and make me proud, they were secretly being gassed like some diabolical plan by Dr. Evil in an Austin Powers movie to use nerve gas to extort billions or millions of dollars from world governments.

So it was game over for a while in the vegetable department. This was sort of all right because I had learned a new passion from my father-in-law, Lorne Archer. Anytime a maple tree seed germinated in his garden he always put it in a little pot. Eventually he had a little nursery in his backyard and was happy to give trees away to anyone who admired them. I had planted a maple tree in my front yard and sure enough after a year or two the seedlings were starting fast and furious. I made the rounds on garbage night and picked up lots of the 6- or 8-inch plastic pots that people were throwing out and started my own little nursery. I took those black walnuts that were bombarding my kids on their swing set in the fall and put them in pots. Turns out black walnut wood is highly desirable and I was starting seedlings by the truckload. Eventually this became the basis for my long-term retirement plans. Since I was never going to make enough money to have a proper retirement, I would find a bit of property, start a black walnut plantation, then 50 or 60 years down the road sell off lumber to pay the taxes.

I didn't stop at these trees though. I grew all kinds of trees. I found an area under a power line where thousands of pine and spruce seedlings were growing. Since the power company was just going to nuke them with an herbicide eventually anyway I grabbed them. Soon my tiny lot was overflowing with seedlings and small trees. One year we had a garage sale and put a few trees out and sold them, which was fantastic. Toward the end of the day a neighbor came over and asked about the pines and spruce. He wanted to know how many I had. "I don't know, maybe a hundred. How many do you want?" "All of them," was his response. It was as if I'd died and gone to heaven! Someone appreciated my trees. Turns out he owns a fishing camp in the north and wanted to grow some natural windbreaks between his cabins. I can't remember what I sold them for. The soil was mostly compost I made for free from leaves that I scrounged from neighbors. The pots were free. The hundreds of hours of my time that I devoted to nurturing them, well you couldn't count that. "How about $2 each?" He paid me $200 cash and it was one of the best days of my life. Oh sure, I was running a successful electronic publishing business. That was good. And the birth of my daughters, oh sure, those were red-letter days. And since Michelle is editing this, my wedding day, now that was the ultimate red-letter day. But selling the fruits of my gardening efforts was off the charts. It's amazing to grow something healthy and beautiful and then share that with someone else. If that person wants to compensate you, it's all the better.

While I was on my tree-growing tear I resorted to growing a few vegetables in the front yard. I discretely planted a tomato and a pepper plant in with our flowers, and no one seemed to notice. The flower gardens were getting bigger and bigger though, so the next year I added a few bean plants. Next thing I knew, I had a vegetable garden growing in my front lawn. It was great. We ran our business out of the house and our customers who dropped in to pick up their artwork thought it was so quaint that we had vegetables in our front yard. Quaint and yummy when a handful of tomatoes were included with every corporate annual report layout. I had also planted a peach tree, which thrived, close to the house, and since we were near Lake Ontario, which regulated the temperatures, it was able to stand the winters. One year we picked a hundred wonderful, tasty, juicy, organic peaches.

I used to work like a dog on weekends composting leaves, weeding, watering, and making the garden bigger. I can remember sprawling exhausted on the final plot of grass we had left, late one Saturday afternoon, and a car with two elderly ladies stopped to make sure I hadn't dropped

My front yard garden with my peach tree to the left and hard-to-see vegetables on the far side of the sidewalk.

dead of a heart attack. I think it was that and the infamous chiropractor incident that convinced me it was time to move. Each year as the garden expanded so did its selection of crops. One year I could wait no longer and the corn seeds went in. Planting corn in your front yard is a crossing of the Rubicon of sorts. It takes you to that next level, from a hobby gardener to a, well, lunatic. It takes you from the interesting character the neighbors humor to a crazy man the neighbors worry will soon be growing dreadlocks and gardening naked while smoking huge volumes of pot.

But step over the line of social etiquette into the realm of eccentric I did, and put in those corn seeds. And they grew tall and straight and so did my pride in my gardening prowess. In fact everything I've learned since says they shouldn't have produced any corn since you should have three or four rows so that the wind can circulate the pollen properly.

Then one day as my chiropractor was contorting my neck and back into positions that they were never meant to attain, he asked if that was corn he saw growing in my front yard. Now I had a reputation. Now I was the guy with corn in the front yard of his city house. Now it was time I got outta Dodge.

Getting out was something that had been in the works for a number of years. We had been looking for a rural property and trying to figure out how to do it. We looked for about five years using the 15-point guideline we'd developed for what our perfect rural home would include. Along with a solar-powered house and large forested area, it would include lots of room for growing food and ideally would allow the land to be certified organic, meaning that it hadn't been treated with pesticides or herbicides for at least three years. For a number of years we had been running an organic produce coop out of our garage. Ted Thorpe, a local organic farmer, delivered a truckload of vegetables to our garage and told us what he'd have the following week. Sixty families would pick up their vegetables, give us money, and order for the following week. Michelle did all the coordinating, ordering, counting, organizing, paying. It was a huge amount of work, but it was one of the best things we ever did to honor our growing environmental concerns. Ted grew sustainable, healthy, organic vegetables less than fifty miles from our home. The families loved the food, and Ted loved having a ready source of income that he could count on every week.

So when we finally found our place in the country and it had never been sprayed at all that we could determine, we were "over the moon" with excitement. The first spring we put the garden where Jean, the previous owner, had put it, behind the house in a very sandy location. We weren't living there full-time so it was only tended on weekends and suffered for lack of water and attention. That year we had an extremely early frost in the middle of September, and when we went up the next weekend the beans and much of what was left was nipped. Moving to an off-grid house we had lots of other priorities, so we didn't pay too much attend to the dismal failure that was the garden. I realized later on that the original garden was in one of the lowest places near the house. These are much more vulnerable to frost, so I wasn't sure if that was the best place to grow long term.

When Michelle decided to start putting in flower gardens she asked me if I could find any topsoil. I found a good spot near the old barn foundation and went back to enlarging the garden in the sandy area behind the house. I kept using my shovel to turn over huge areas and then raking the grass out to leave what soil I could behind. It was very time-consuming and tiring, but I loved it. This was the garden I was going to be growing food in until I dropped dead in my potato patch. So I didn't mind the work. The first full summer we were there I planted a large garden in this sand and I think it suffered from the Neonatal Intensive Care Unit term "Failure to Thrive."

Michelle and our friend Ellen try to identify herbs in our original garden left over from the previous owners.

I did continue to retrieve wheelbarrows full of much nicer soil from near the barn foundation for flower gardens and to put around apple trees we planted. I'm not sure when it dawned on me, but eventually a high-efficiency compact-fluorescent light bulb (since we were off the grid) went on over my head: maybe I should actually have my vegetable garden there. I believe the expression that comes to mind regarding my delayed realization is that I'm obviously "not the sharpest tool in the shed." In this matter I do not disagree.

I guess what held me back was the fact that the land around the old barn had four-foot high grass with a tangled, thick network of roots that made removing it a huge hassle. An intelligent person would have looked at that grass and said, "Man, there's gotta be something great nurturing this growth." I looked at it and said, "Man, it's gonna be lot of work turning it into a garden." While my assumption was correct about the work involved, it turns out it was worth it.

I started with a small garden in that area and kept the one out back in the sand. Each year I hacked off more grass and made the new garden bigger. The soil there was much healthier. As much as I enjoy the physical effort of enlarging a garden with a shovel and claw-like cultivator to shake the soil out of the grass clumps, I realized there had to be a better way. Since I didn't own a tractor to plow it with, and since my rototiller just rolled over and played dead every time I tried to use it, I needed another strategy.

Eventually I realized that many of the farmers around me often ended up with large round hay bales that were not fit for animal feed. So I started purchasing them and using them to kill the grass. I'd roll out a thick layer of hay over the area I wanted to be garden next year, and after about six months the grass underneath would be dead. Then I could rototill the hay into the ground underneath with the added bonus of adding more organic matter to the new area. It worked like a charm, and ever since I discovered this strategy my garden has grown exponentially.

It's now getting close to half an acre, which when you own 150 acres doesn't sound like much, but which when you look at a garden full of weeds that need pulling is in fact a very large garden. As we got to know the various previous owners of the house, which dates back to 1888, we learned the history of the area where the garden is. We discovered that animals had been kept in the area around the barn and much of the manure removed from the stalls was spread in this area. We also found out from a woman now in her 70s that they had a garden in the same place when she was a child. There's a lot of history in our garden.

The barn was torn down several decades ago, but the concrete foundation still remained. Trees had grown in all around it and it was in full shade. I started cutting the trees back to let more sunlight in. Inside the barn foundation itself was a jungle of sumacs and other small woody trees and plants. I started hacking away at these and discovered that there was actually a concrete floor in the barn. Once I got the overgrowth removed I was able to move all the soil that was left into some raised beds. I made the raised beds with cedar posts I scrounged from the side of the road as the township fixed the wire guardrails. So I now have four raised beds in the foundation. The soil in them is wonderfully dark and rich. I put my heat-loving plants in there, like peppers, eggplants, and tomatoes. This year I tried sweet potatoes there as well as peanuts and okra. We can still have cool evenings well past our last frost date, and the foundation is a great heat sink which absorbs heat from the sun all day and radiates it back at night. I think that even though there is no roof, the walls help cut down on some of the cooler winds that can retard growth early in the season.

My biggest challenge with gardening here is water. The good news about a very sandy soil like mine is that it has great drainage and you don't have to worry about water sitting around and causing problems.

The barn foundation garden.

The bad news is that it dries out quickly, and we seem to be in a very drought-prone area. Near the barn foundation at one end of the garden is a dug well. This is a concrete-lined well that was built in the 1920s to provide water to the animals. It had an old building that was crumbling around it and a wooden pump that had fallen down into the well itself. Once I pulled down the building and cleaned out the well, I had an excellent source of water near the garden. Eventually I built a solar-powered pump that allows me to move the water to where I need it, either to rain barrels for hand watering or into my irrigation system.

Every once in a while I will think I have enough garden under cultivation. You can grow a lot of food on half an acre. We have a plant-based diet, which means that we can actually offset a fairly large chunk of our food budget each year with the garden. But then I'll see a logical spot near the garden where the soil is pretty good. And think well maybe next year I'll be selling vegetables in town, so maybe I should expand the garden just a little. I went from planting 4,000 heads of garlic last year to more than 10,000 this year. It grows well here and since I plant it in the fall and the bulbs are large enough in the spring that the cutworms can't usually damage them. So the temptation to expand is always there.

I continue to try new things. Two years ago I planted several apple trees in a new raised bed I put in the barn foundation. I put in more

strawberries and raspberries last year. While I struggle with cutworms to keep these thriving, last year we had a great crop of raspberries. They grew well everywhere and we froze a huge amount of the ones we couldn't eat.

Last year we put in our first blueberries. These are the high bush kind that are easy to pick and which produce large berries. Blueberries take a long time to mature. We'll get a few from the ones we planted last year, but it will be three or four years I think before we get a large harvest. But that's the beauty of having a place like this where you put down roots, in this case blueberry roots. I don't mind taking the time because I know I'm going to be here for as long as it takes. It's like steel roofs. Steel roofs last forever, but they cost a lot more than asphalt shingles. I think a lot of people have trouble rationalizing the additional expense, because they don't see themselves living in the house long enough to recoup the investment. As we start down the backside of the peak oil curve, I think a lot of people with asphalt shingles, made from a barrel of oil are going to regret that decision.

This spring I invested $200 in 21 new blueberry bushes. I got seven each of three different varieties. These were two-year-old plants that came in pots as opposed to bare root. They are big, healthy plants. And they look amazing. I planted them in an area under the wind turbine where the soil is very poor. Well, it's just sandbox sand basically. But I put in manure and worked in rotten hay. I used some topsoil around the plants and I have been dumping coffee in this area. Pounds of coffee grounds. In fact, tons of coffee grounds. Blueberries like an acidic soil and coffee grounds are acidic. It turns out that Starbucks will give you their used coffee grounds for free. So every time I go to Kingston I come home with 200 pounds of coffee grounds. The car smells great and it makes it tough for Michelle and I to give up coffee let me tell you. But the soil is starting to really look better. Having an area of the property that I can nurture and bring along like this really is a dream come true for me.

I have some things that I don't put as much effort into. We never seem to make much use of the rhubarb, so I haven't been doing anything to keep it thriving. Luckily it's a perennial that comes back every year, even if you completely ignore it. I put manure on it every fall and still have huge plants.

Having more room to expand my garden is a realization of a life long passion for me after gardening in such restricted areas for so long. This year I'll celebrate my 52nd birthday, and since my first garden was the year I got my driver's license at 16, that makes it 36 years that I've been gardening. It's something that is very much in my soul. Somehow it was in my DNA. And every year is the same. By the time I've put the last squash in the root cellar in the fall I think, "That's it, I've had enough. It's too much work, I'm not having a garden next year." And I'm serious. And then after a month or two of the snow I start to get a low level longing for green plants and brown soil. In February we start going through seed catalogs. In March the sweet potatoes are sending off shoots so we start putting them in small pots. Then we start a few trays with vegetables. And by April I can't wait to get out and start planting again. The urge is overpowering.

In a world where our diet is so dependent on fossil fuel, and the International Energy Agency has now finally admitted we hit peak oil in about 2006, growing your own food is going to become increasingly important. I know now I can grow all my own food. Our diet would be pretty bland, but I can grow enough grain, and corn and potatoes to sustain us. This is a very good feeling. A very comforting feeling.

Every growing season brings its own unique challenge. Too much rain. Not enough rain. Too many cutworms. Not enough…. But that's what growing food is about. It tests your stamina. It tests your will to reap the fruits of your labor. And sitting down to a meal that you've just

brought from the garden makes all the effort worthwhile. Cooking it with electricity you made from the sun and wind completes the circle and makes it feel like a very "whole" process. And for the gift of being able to live this way, I am eternally grateful.

A day in the life of Sunflower Farm

We've described many of the various activities that take place around Sunflower Farm but you're probably wondering how they all fit in to an average day around here.

Since I'm writing this in July, I'll start with a description of a typical summer day here. We get up early for two reasons. First off we wake up early in order to get out in the garden and get as much done as we can before it gets too hot. Secondly, since the sun is up so early and shining in our windows, it's hard to sleep in. Today I was awake before 5 a.m., and I got up shortly after. I went right outside to relieve Morgan the Wonder Dog of his "Deer Watch" Duty. Morgan has been sleeping in the strawberry patch lately since a small deer has decided to include our strawberry plants in her diet. She'd eaten about 1/3 of the patch before I noticed, so I've been tying Morgan up in the patch to guard the strawberry plants at night. The strawberry harvest is over for the year, but I need to water and nurture them all summer so that we'll get a good crop next year.

Then I let the chickens out of their coop and they fight to be the first one to ride the gangplank drawbridge door to freedom. They get fed and I clean out their coop and put in fresh wood shavings. Then I water the cows (two mothers and their calves that belong to my neighbor) and feed them fresh hay.

My next 3 to 4 hours are dependent on the weather. If we've had rain, I won't have to spend too much time watering, but the rain will have inspired the weeds to germinate so I'll spend my time in the garden hoeing and weeding. If we haven't had rain, it means less time weeding and more time watering. I also continue to plant new rows of lettuce, brassicas like broccoli, beans and other vegetables so that I can harvest them all summer long.

We are now experiencing a drought so basically all I do is water. I use the hose to water the berries around the house. I can cycle the pump on our drilled well about 4 times comfortably in the morning before I've surpassed the refresh rate. In between these I use watering cans to target specific areas, like the sweet potatoes in the new garden behind the house.

As soon as the sun is up high enough I make sure that the solar pump in the main garden is on. I usually start by filling up the rain barrels I have strategically placed around the garden, which I use to fill up watering cans and water specific rows by hand. I water one area very thoroughly one day, and then generally leave it for a couple of days before I water it again. I want the plants to get a good soaking so I make sure that the water gets deep and then I leave it for a few days. This forces the roots to go down deep to find that water which makes them more resilient to the drought and will also help them to find more of those wonderful trace minerals and elements in the soil that I want them to absorb for us to eat.

By 9 or 10 a.m. it's starting to get really hot, so I have breakfast. Then I head to the office for "office work" which right now is putting the finishing touches on *Little House Off The Grid* to get to the editor. My work in the office also involves various things like blogging, responding to emails and burning DVDs.

About once an hour I head to the garden to move the solar pump hose. After the rain barrels are filled I put the hose into one of the two drip irrigation systems I have on the go so it will work away without me having to monitor it. With our sandy soil, drip irrigation is a wonderful way to have the water taken up by the capillary action of the soil, which draws the water to form a fairly wide area under each of the drip heads.

Three or four times a day I turn the solar panels to track the sun. This is particularly important at this time of the year because the sun has such a wide path. During the day I also periodically water the berry patches near the house with the hose, leaving more and more time in between watering to give the well plenty of time to refresh.

There are usually some pressing outside jobs that I fit in amongst my office work as well, such as squishing scarabs on blueberry plants, building more drying racks for garlic, fixing "X" or "Y" piece of equipment/machine/computer/watering system. These are nice tasks because they break up my time in the office. My back appreciates the time spent away from my computer desk.

After dinner it's back to the garden for maintenance work like weeding or more targeted watering with the watering can. In the next few weeks as we start harvesting garlic, evenings will consist of hours on the front porch cleaning garlic to sell.

I like to try and watch the CBC National News at 9 pm just to see the sort of madness going on the world on any given day, but at this time of the year it's a lost cause. I have trouble holding my eyes open, so I usually just check the cows' food and water, coach Morgan on the importance of his Deer Watch Duties and sneak up on the chickens who have put themselves to bed in the coop, and close the door before they storm out thinking I've brought them some treat to eat.

During droughts like this one, life starts to feel like a real grind. I appreciate that weeds aren't germinating as fast with the lack of water, but the further into a drought we get the more moisture-depleted my sandy soil gets which is kind of depressing. After a wet period my soil looks dark and rich with lots of organic material from all of the rotten hay and manure I've added over the years. During a drought it looks tired and sandy and no amount of watering I do can replace what I need nature to do periodically, which is to give everything including the lawn (a.k.a. possible future gardens) a good soaking.

I have grown more and more resigned to living in a drought zone. I spend less time raging against the skies at their lack of rain clouds and more and more time focused on storing water when it does rain, and finding other ways to efficiently move and dispense water to where I need it in the garden.

During the summer I fantasize about December when my bedroom will be cold and dark and I can sleep for 10 hours a night to try and make up for some of the sleep that I've lost over the summer months.

I really love the seasons here at Sunflower Farm. I love the summer because I am able to grow so much of my own food, but it's a real grind sometimes and feels more like a marathon that lasts for 3 months of the prime growing season. But as grinding as it is, I wouldn't trade a minute of it for a return to our life in suburbia before we moved to this little peace of paradise.

16 Heating with Wood

by Cam

Oh I'm a lumberjack and I'm OK.
Monty Python

And when I come home cold and tired
It's good to warm my bones beside the fire
Pink Floyd, HOME

There is nothing like wood heat. Nothing warms you like sitting beside a wood stove and feeling the heat radiate out at you. It simply warms you to the bones.

When we lived in the city and had a natural gas furnace, our home never felt very warm. Now granted, we did our best to keep our thermostat low to use less energy and reduce our footprint on the planet. But when you have small children playing on the floor, you tend to keep a home warmer than you might for yourself, yet we never had that warmed through feeling.

Our house in Burlington did have a fireplace, but it had the opposite effect you'd expect from a fire. While it added to the atmosphere tremendously, it acted like a huge siphon, pulling warm air from the house letting it go right up the chimney. It was a double whammy for us, because the thermostat was located right across from the fireplace. So as the fireplace managed to heat up some of the air in the living room, the thermostat was happy and wouldn't bother turning on the furnace. So by the time we were ready for bed, it was like winter camping. We added fancy glass doors to the fireplace, but it was still an inefficient, leaky, disappointing exercise.

When we bought Sunflower Farm, we knew we were going to have to heat with wood. At first, this was a bit of a concern. People in the city warned us "Oh that's so much work, you'll never be able to do it." And what about wood smoke? Wasn't it bad for the planet? So we had some reservations, but luckily we were used to spending Christmas at the cottage, where shorts and T-shirts were the usual festive attire, as Dad kept the cottage at tropical temperatures, while making hourly announcements of the outside temperature, often significantly below zero. So we already had a pretty good exposure to heating with wood.

Many off-grid homes that haven't been built from scratch using advanced passive solar design, will use wood as the source of heat. This simply is because many will not have sufficient power to run other sources of heat. Obviously electric baseboard heat would not work. Even running a furnace fan can be a challenge, because the time of the year you need it most, is the time when you tend to have the least amount of power available to you. Someone living in an off-grid home might look at using a ground source heat pump. For an on-grid home, faced with skyrocketing natural gas prices, this is an excellent option. However, off the grid you still face the challenge of needing it the most at a time of year when your power production is limited. The heat pump requires moving a liquid from inside the home to outside the home and circulating it through the ground to absorb some of the heat and bring it back into the house. Then you must run a heat exchanger to capture that energy. Moving a liquid like this is a double-edged sword. If you have a stream on your property, capturing the energy of the water's movement offers tremendous potential electrical power, but doing the opposite and having to pump water over long distances requires more power than most off-grid homes will readily have available. Ground source heat pumps extract 3 units of heating (or cooling in the summer) energy from the ground for every unit of electricity that goes into running them, but you

simply won't have that in an off grid home.

So wood heat becomes the most attractive alternative. Most off-grid homes are usually in a rural area, with easy access to firewood. In our case, all our firewood has been sustainably harvested from our own property.

The woodstove the previous owners had installed was a Vermont Casting "Defiant Encore." Anyone who saw the house and knew woodstoves would say "Oh, a Defiant Encore! That's the Rolls Royce of woodstoves!" This set high expectations, and as it turned out, the woodstove was closer to a Ford Pinto, after it had been rear-ended. It was a train wreck, but we just didn't know it yet.

Wood heat is the only carbon neutral way to heat. As a tree is growing, it is using photosynthesis to convert CO_2 into its woody material, while giving off oxygen. Storing carbon dioxide is a very important thing in a world where we have too much being pumped into the atmosphere. When that tree dies and falls to the forest floor, it is going to gradually release that CO_2 back to the atmosphere as it decays. It is also going to create heat as it rots and returns to the soil. In fact, that tree will release the same amount of CO_2 and heat if it died and rotted in the forest, as it would if I brought it into our house and burned it. The difference is I'm accelerating the process. That's why it's called carbon neutral. The CO_2 that is released when you burn it is what the tree had captured from the atmosphere while it was growing.

If you heat with oil or natural gas or other hydrocarbons, you are taking carbon that was created millions of years ago as plant materials trapped under layers of water or other debris, and releasing it. You are taking carbon that is safely locked away below the ground, and by burning it; you are releasing it to the atmosphere. If you heat with electricity in an area where the electricity is produced with coal, you are releasing massive amounts of carbon dioxide through your use of electricity. Coal is basically pure carbon, so burning it releases an inordinate amount of carbon dioxide into the air. Fifty percent of the electricity generated in the United States comes from coal-fired power generating plants. And even if you do the responsible thing and install a ground source heat pump, while it will reduce your carbon footprint greatly over heating sources, if the electricity that powers your pumps and heat exchanger comes from coal, your system is not carbon neutral.

The key with wood is to burn it very efficiently. This has been the goal of back-to-the-landers since the 1970s when North America experienced its first energy shock. Reading an issue of *"Harrowsmith Magazine"* from the 70's you'll see dozens of woodstove manufacturers, all tweaking their designs in pursuit of a clean burn that maximizes the heat and minimizes the environmental impact. One of the designs that became popular is one that uses a catalytic combustor. In a woodstove like the Defiant Encore, you burn your wood in the main combustion chamber that you can see through the glass doors. Then when it's burning well, you put it into "airtight" mode. This slows down the air getting to the fire, and redirects the smoke through a baffle at the back where it has to pass through the combustor. This is a piece of refractory material that resembles a honeycomb that gets very hot. So hot that when the smoke passes through the combustor, any unburnt gases and particulate materials are combusted. What is left is basically water vapour, and an efficient, clean burning, EPA (Environmental Protection Agency) certified wood stove should be operating close to the efficiency of a natural gas furnace.

Like so many of the systems at Sunflower Farm, getting our woodstove to work well was a major challenge. One of the keys to an "air-tight" stove is only allowing air in where and when you want to. When you're starting it, you want lots of air getting in. When you put it into "airtight" mode, you want very little air getting in. This allows the wood to burn slower. When you go to bed at night, you load it up get it roaring, put it into airtight mode and hope it will burn through the night. Keeping the air out involved replacing gaskets in a number of places, such as around the front doors, ash pan door and the cast iron door on the top, which was convenient for cooking on, but was another place for unwanted air to get in.

Our first winter we struggled and battled with this stove, but it won every round. It refused to burn properly, and while it would have been nice to have the proper 8 hours of sleep and get up to a woodstove with glowing embers we could place new wood on, this was a recipe for a cold house. The stove would not go 8 hours, and we'd be lucky if it went 5 or 6. That first winter we were still keeners, and rationalized it that we'd have to get up to go pee at 2 or 3 in the morning, so we might as well stoke the woodstove. There is a big difference between dragging yourself half asleep to the john and back, versus having to lift firewood into a woodstove and wait until it starts to burn well. After that much activity it was usually tough to get back to sleep.

This stove also had a tendency to overfire. That meant if when you loaded it at night, if you let the wood get too far along in its firing, even when you shut it down and tried to restrict the airflow, the stove had a mind of its own, and it would just keeping burning away furiously, as the stove pipe at the back that took the exhaust gases up the chimney, glowed red hot. It resembled a coke oven at a steel mill and whenever it happened I felt like I should fire up Orf's "Carmine Borana" which is so often used in movies whenever the devil is about to make an appearance. This was terrifying and you had to open the doors and sit there for an hour while it burned itself out. By then you were so tired you basically went to bed without putting in more wood, and got up to a very cold house.

The stove also liked to let some of that smoke back into the air out of the various holes that were supposed be sealed. This meant that as you lay there your nose would catch brief whiffs of smoke smell. I think it would be safe to say that it is in a homo sapiens' DNA that the smell of wood smoke can be a potential hazard that requires examination. If a forest fire was headed your way, it might be a good idea to get a move on. And so it was that first winter that we had to steel ourselves to the fact that the house was not in fact burning down, it was just that "Rolls Royce" of woodstoves downstairs back drafting.

Before the second winter hit, I decided it was time to replace the catalytic combustor. The refractory material broke down over time and needed to be replaced. When I pulled the back of the woodstove off to get to the combustor, I was perplexed with the positioning of the combustor. The honeycomb holes were facing me as I pulled off the combustor, but it seemed that the air that had been diverted to the back of the woodstove should actually start below and move up through it, as hot air naturally rises. The way it was positioned in fact restricted the airflow. A quick call to the woodstove company that sold us the new combustor proved my theory correct. We had spent the first winter with the combustor going the wrong direction, restricting the air's natural movement and causing the stove to function very poorly. The lesson learned was that if you inherit a woodstove, get a professional to make sure it's working properly. My assumption that everything was as it should be, was wrong.

The second winter, even with a new combustor positioned in the correct way, and with the many gaskets sealed as well as we could manage, the stove only worked marginally better. It still back drafted releasing some smoke into the house, and it would not burn anywhere close to 8 hours. Oh well, another winter without sleeping through the night was a small price to pay for living in paradise.

After that second winter we decided to get the stove overhauled. We hired a local woodstove expert to help. He spent the day; basically tear-ing the stove apart and putting it back together meticulously, ensuring everything fit properly and that the gaskets were installed properly. Gary also trained us on how to properly clean the chimney, an important part of life with a woodstove. The stove was made mostly of cast iron, considered to be a good material for woodstoves, because it is able to withstand the stress of the heat and cooling cycles. The challenge is that cast iron can't be welded, so it is held together with bolts and cement. With the constant warming and cooling, expanding and contracting, eventually gaps open where you don't want them and unwanted air gets in. After it had Gary's stamp of approval, he pointed out a tiny crack in the back plate of the stove and said it shouldn't be a problem, but we should watch it.

Well that third winter we did watch it. We watched that piece of metal turn into a twisted wreck of its former self, looking like those photos you've seen of two train engines running into each other. So by the spring, we had decided, the Vermont Castings had to go. It was during this time that'd we'd come to know Dan Creighton who worked at a shop called "Renewable Energy of Plum Hollow." They had started in the 80's, trying to eke out a living selling solar and wind power. Eventually they had decided they needed to supplement their income, and now also sold woodstoves. They sold a woodstove from a company called Pacific Energy. Unlike the catalytic combustor technology, this non-catalytic woodstove used a second oxygen burn, bringing in oxygen to burn any remaining gases or particulate before it went up the chimney. The stove was much less complicated, had few moving parts and had just one gasket that ran around the large, glass front door. It was steel, so it could be welded together to ensure it was airtight, and was lined with bricks inside. It also didn't have the on-going expense of replacing the catalytic combustor.

In the guesthouse we had a Dutchwest woodstove, which also had a catalytic combustor. While we had been told that a catalytic combustor should last 3 to 5 years, we were spending $300 every second year to replace them because they simply crumbled and were of no use as an emission control device in their deteriorated state.

While we should have researched the purchase of the new woodstove for the house thoroughly we didn't. We knew and trusted Dan, and he just kept saying, "This is the stove you need." And he was right. The stove is absolutely fantastic. It works like a dream. We can sometimes get it to burn for 12 hours. We would have been happy with an 8-hour burn after the previous woodstove. But with this, you could leave the house for the day, and come back 12 hours later, toss in a few logs on the glowing embers and it would be roaring again in minutes. The stove is also made of regular steel and is lined with bricks to reflect the heat away from the walls. Because it's made of traditional steel, it can be welded together which ensures an airtight fit. Cast iron cannot be readily welded, which means parts are bolted together, making it easier for air to get in. And because it burns so efficiently, we are burning less wood. As with so many things, we wish we'd done it the day we moved in. But it takes time to develop the knowledge and network of experts who can help you make the right choice.

Now we always recommend people go with this design. Make sure it is EPA approved, which means that Environmental Protection Agency has certified that it burns cleanly and removes the maximum amount of particulate to ensure you minimize your impact from heating with wood. And go with a non-catalytic design. They minimize the amount of air that get into the woodstove and maximize the efficiency of the burn.

Our house is well set up for wood heat. When Jean and Gary renovated the old farmhouse, they took previous small rooms and opened them up. This allows the heat to move around. The woodstove is located in the living room at the bottom of the stairs so the heat rises to bedrooms on

Our new Pacific Energy woodstove loaded up with pots of water for tonight's bath.

the second floor. This is really important in an off-grid home. A typical furnace uses electricity for the fans, which move the heat around. By correctly positioning a woodstove, the heat will move around the house without the need for electricity.

The other element of wood heat, is procuring the fuel. You can usually purchase firewood in a rural area, but this obviously increases your requirement for an income. So far we have cut all our own wood. So far we have not had to cut a single living tree. For the first 7 or 8 years we burned oak, an exceptional hardwood that provides amazing heat. A few years before we bought the place, there had been in infestation of gypsy moths in our area, which killed many of the oaks. So now it was simply a matter of cutting them down and turning them into firewood.

The root systems of many of the oaks rotted out once they died, because they were in contact with the wet soil and more susceptible to more rapid decay than the upper parts of the tree. This was nice because it meant that most of the oak I cut had blown down in windstorms and didn't need to be felled. As I've gradually cut the dead oaks, the beavers have filled the gap and become our partners in wood harvesting. They have a voracious appetite and cut much of the standing trees around the ponds on our property. The silly beavers of course, only want the small branches and tender wood to eat. So they cut these mighty trees, trim off the small stuff that I would otherwise not use, and leave the firewood that I need. They also take the risk of felling the trees. It's a symbiotic relationship and while I wish they wouldn't cut back as far they do from the ponds, as long as they're going to cut down trees, I'm not going to let them go to waste.

We have 150 acres that is mostly treed, and this is more than enough land. In fact, there is much literature that says you can heat a home comfortably with 20 acres or less. Much will depend on the type of trees, and how quickly they regrow, but a properly managed forest can be an extremely responsible way to heat your home. The key is to harvest mature trees that as they fall, increase the sunlight reaching lower trees, allowing them to grow faster and be ready for their eventual cutting.

Cutting firewood is something I loved doing from the moment I started. It's physical work, so it naturally creates endorphins that give you a buzz. Afterwards, although you can be exhausted, there is always exhilaration. It's compounded by the fact that you've actually accomplished something. It's not like riding a stationary bike or using a weight machine. When you've cut wood for a morning, you have a pile of wood to show for it. There, look at that, I did that! As I spent more time doing it, I got better at estimating how much wood I'd cut. I could look at a freshly split pile of oak and say "That's about two weeks worth of heat in the coldest part of winter." That is a really cool experience! Well, actually, it's a very warming experience.

There is an expression that "Wood warms you twice, first when you cut it, then when you burn it." I've modified it slightly to "Wood warms me about 7 or 8 times." First I fell the tree, then I cut it into burnable lengths, then I have to get those to the road where I can get them in the truck, then I have to split them, then I have to load them in the truck, then I load them in the woodshed, then in the winter I take them from the woodshed into the house. It's about 8 separate and distinct tasks that might seem hopelessly inefficient, which I absolutely love. I love every step along the way.

The trees that didn't fall on their own or with the help of the beavers, I've learned to fell myself. Nothing is more exhilarating than chopping down a tree with your chainsaw. If you calculate the natural lean of the tree, and make your cut properly, it should fall exactly where you want. After 10 years I believe I have become very proficient at it, even though there have been many trees that did not go as planned. Luckily though I learned the rule of having a cleared escape path. There were many times

when this came in handy. But being a city boy, I tended to err on the side of caution.

Once the tree was down and cut into burnable lengths, I had to find a way to get them to the road where I could get them with the truck. Our property has an old road, which runs through most of it. In the 1940s, the road to town in our neck of the woods was very rough. On part of our land, there is a bog, where there was a "floating bridge." This was a series of large logs, tied together with heaving wire. So, you literally drove your team of horses or vehicle, over the bridge that floated on the bog.

Eventually the township had enough of vehicles tumbling off, and the owners of our property deeded them a detour through the property. When the new road went through in the 1980s, the old road was abandoned, and it is now a perfect place for us to walk and get the truck in for firewood. Our property is fairly long and not that wide, so it makes it easy to get wood out to the old road. Many people in the woods now insist on using an ATV to haul firewood. I don't want to spend the money to buy an ATV, to burn the gas, have one more machine to break, to pollute the air and detract from the peace and quiet, or to have to make new roads to drive the machine through. Instead, I drag the logs out with a sled. You can buy heavy-duty plastic sleds designed to be pulled behind a snow mobile. So once we get enough snow, I start pulling the logs out to the road.

I get into a routine with this. After a fresh snow, I'll pull a few loads, and then leave the track to harden and pack down in the cold. If there are gulleys, I'll fill them up with snow, and I've even made some paths up the side of hills after a deep snow, where I pack the snow down like you would build a road into a mountain side. Once the track is hardened up, it makes pulling the wood much easier. That's not to say it's easy, but much easier than dragging the wood with no snow. It's a slow process and exhausting, but there is no better activity for meditation on the planet. Once you get into the groove and have walked the path a number of times, you know the places where the sled will slide forward and catch up to you, and you know where you'll have to pull harder. The topography of our land is very hilly, and a lot of the time I'm pulling wood out of a valley because that's where the water with the beaver ponds is. Hauling wood like this allows me to switch off my mind, and get into a rhythm. It's quiet, I'm in fantastically beautiful woods, I see wildlife, and I become very much one with nature. The only fossil fuel involved went into the diesel tractor that grew the oats for the granola that fuels me.

I'm not sure I completely understand the concept of Zen, but I believe there is nothing closer to a Zen-like state, than pulling a sled full of firewood on the snow in my woods in paradise.

The wood is going to warm our house, producing the same amount of heat if we left it to rot. And it very much gives me an incentive to get out of the office. I get into a routine with the seasons. Spring is for planting the garden. Summer is for weeding and watering. Fall is for harvesting, and by late fall as it's getting colder, I get into the routine of firewood. As the days get shorter, I'll often stop working on the computer around 4 pm and head out into the woods. After an hour or two of harvesting firewood I've had some great exercise, and feel absolutely blissed out. I'll usually go back to work in the office for a few hours after dinner which I don't mind in the least, because I've used daylight hours for outside work. It's an absolutely amazing lifestyle, not one I had intended on having, but one that has evolved into a wonderful way of life.

The health benefits, both mental and physical, are amazing. I have remained between 140 and 145 pounds my entire adult life. People have often said, "It must be so great to have your metabolism!" Yes, I believe I have been blessed with a great metabolism. But I have also spent my entire life getting lots of exercise. For the last twenty years, I've eaten a vegetarian diet, low in animal fats, and high in fruits, vegetables, grains and starches. And for the last 10 years, I've worked my ass off. I sweat like a dog all summer planting, weeding and watering the garden, and as soon as that's finished I go into firewood mode for the winter. There are some times I'm sure I'm burning more calories than I'm taking in. Those times usually catch up to me, and I'll back off the work at bit, and maybe eat a few more eggs, from our happy chickens.

But my success in keeping my weight consistent has been 10 percent inspiration and 90 percent perspiration. I realize everyone can't heat with wood, but for us, it's a blessing. I did a rough calculation last year, keeping track of expenses for us to heat with wood. I did not include the cost of the woodstove, because most people only calculate their fuel costs. If I add up all the gas and oil bought for the chainsaw, and include one tank of gas for the truck to haul it from the bush to the woodshed, I spent about $125 to heat our home last year. If I include the $800 cost of the chainsaw, which I'm sure I'll get 8 years use out of, it makes it about $225.

Now if I include the opportunity cost of what I could make if I put all those hours of wood cutting and hauling into productive income earning effort, it's probably $3,000 that it costs us. But I love doing it. I also don't include the cost of the health club membership I don't have to buy. Or the cost of gas to drive to the health club, and the trendy exercise clothes. My wood hauling attire would make the shabbiest street person proud!

Burning off excess electricity "bucking" logs into woodstove lengths with the electric chainsaw.

As I've become more aware of the issue of climate change and of peak oil, which will start increasing all heating costs, including mine through the use of gas in the chainsaw, I've started to try and reduce the gas used. Now when I cut wood in the bush, I cut it in 3 or 4 fireplace log lengths. Then I haul these back to the house where I "buck" them into fireplace length with my electric chainsaw. As I've continued to add solar panels, even with my hot water tank diversion load, I have days where I still have too much electricity. So now, using my electric chainsaw helps me reduce not only my carbon footprint, but also my reliance on fossil fuels for the gas chainsaw.

As I get older and splitting the firewood starts to take it's toll I'm going to purchase an electric log splitter that I can also use as a way to use up excess electricity and save some wear and tear on my back. Along those lines we used to carry our firewood from the woodshed into the house by the arm full. Recently I fashioned a wood hauler out of a good quality handcart dolly. It now takes two trips to fill the woodbox versus 7 or 8 before. As I get older I guess I'm looking to work smarter rather than harder.

Wood heat is not for everyone. The conventional wisdom seems to be that you need 20 acres of hardwood or mixed forest to heat with wood sustainably. This means cutting mature trees and allowing smaller ones to get more light and mature in their place. Heating with wood is a lot of work, but it is good work. If you burn it properly in an EPA certified wood stove, it is a very responsible, renewable way to heat. It's not releasing CO_2 that's been safely sequestered underground for eons into the atmosphere, so it's good for the planet. If you have the time and property, it's very good for your pocketbook, your body, and it's very good for your soul.

A WORD FROM OUR EXPERIENCE
Buy a good wood stove that works! Older models that aren't EPA certified and don't live up to their performance potential will detract from the wood burning experience and increase your impact on air quality.

Happiness is a full woodshed or wood pile.

17 Barbarians at the Gate
(Security at Sunflower Farm)

by Cam

Many of our city friends freak out when they think about living in the country. I don't know whether they've watched too many horror movies or watch too much TV news and have a false sense of the likelihood of being the victim of crime, but they seem to think living in the country is inherently dangerous.

Michelle and I have never felt more secure than we do living in this place and haven't really worried about it. We never lock our vehicles in the driveway and we chuckle at our guests when they lock up their vehicles and have to find the keys when they need something from the car. We point out that we are a long way from our nearest neighbors and so it's pretty unlikely that someone is going to show up and steal something out of their vehicle. We often don't lock our doors. Certainly not in the summer when we leave the doors open to allow the house to cool off as much as possible.

I love knowing that our nearest neighbor is 4 kilometers, or 2 ½ miles down the road. I guess I should be freaked out, but I'm not.

This is not to say that there haven't been moments when I have felt less than comfortable with our isolation and so it's only fair that I share those moments with you. We feel very safe and secure in our little piece of paradise, in our cozy little house off the grid, but as Obi Wan Kenobi said in "Star Wars," sometimes there can be minor disturbances in the force.

The first episode occurred shortly after we had bought the place. A woman and her twenty-something-year-old son dropped in for a few visits. It turned out that she had been looking to buy this place and was curious to find out who had managed to do so. The first visit was friendly and she explained that she wasn't able to get the financing together. The second visit was a little disturbing, or at least a bit strange. She was a little bit more forthcoming about her inability to buy the place and acted as if somehow I was responsible for her not having the financial where-with-all to do it. Then she started rambling on about how she had often thought about building a home where the farm animals slept downstairs which would help to heat the upstairs where she planned to sleep. She had some other unique ideas that may have been very logical but which I was not familiar with. As the discussion got more and more "out there" I elected to cut the visit short and to get back to the projects that I was working on. I felt I might have stepped over the line of country friendliness and hoped she'd get the vibe that I didn't enjoy her visits. Apparently she did because we never saw her again. This was our first taste of many unusual "drop in" visitors that we've had over the years.

In the early days of our life here a lost dog showed up at the house and so we called the local animal control officer. When he came to pick the dog up he asked if we had a gun. We told him that we didn't. He suggested that it would be wise for us to get one because you never know when a rabid animal is going to show up. Or if a bear were mauling one of our pets, or worse, one of us, we'd want a gun. I hope he meant to scare it off, since I doubted the person being mauled would be any better off after I'd fired a few rounds at the bear. I have serious concerns about my accuracy in a time of high stress like that. I think he just meant to fire a few rounds into the air to scare the bear.

One day when I was in the city on business Michelle took our dog Morgan for a walk. As she and Morgan came around a corner, on the road not too far from the house, she was shocked to see a black bear and her cub crossing the road. A black bear is one thing, but this bear was obviously a mother with a cub in tow. While you might be able to reason with a bear, a mother who felt you were a danger to her cub would be another story. I think I agree with Sarah Palin in this area. So shortly afterwards we bought a can of "bear spray." This is basically pepper spray and it hangs by the front door. When we bought it from a hiking store in downtown Kingston they took our name and contact information, sort of like we were about to commit a crime. We have yet to use the bear spray and hope we don't ever have to.

Sometimes when we go on walks in the spring and I figure the bears will be waking up hungry from a long winter's hibernation, I remember to attach the bear spray to my belt. This of course is completely illogical since I spend a great deal of time in the woods and rarely take the bear spray. And we all know that when I DO encounter a bear I will NOT have the spray and will have to use all the other techniques I've read about in order to ensure that my bear sighting ends well. This will involve raising my arms and making my hulking 145-pound frame seem large and intimidating to the 400-pound bear. I will also do some tough talking to the bear, and slowly back away without looking the bear right in the eye to challenge him or her. This is, of course, the textbook technique to avoid a bad outcome. How likely I am to remember all of this and behave appropriately is another story completely. Personally I hope I will not be alone during the encounter. As my neighbor Ken says, "You don't have to be a fast runner in an encounter with a bear, you just have to be able to run faster than the person you're with."

One day we had an Ontario Provincial Police (O.P.P.) officer stop at the house while he was investigating something, the details of which I can't recall. The O.P.P. are responsible for policing this rural area of the province. He made a comment about our location and suggested that we are about 30 minutes from the Sharbot Lake detachment and 30 minutes away from the Napanee detachment and sometimes when a call comes in from this area it is kind of a toss up as to which detachment would respond. I kind of got the feeling it would be one of those "You go. No, you go" kind of discussions as to who would get stuck coming out this far. He suggested that basically we are on our own for 20 or 30 minutes anyway unless by some fluke there is a car nearby. Since I can count on one hand the number of times in 14 years we've seen an O.P.P. cruiser on our road, this kind of stuck with me. It was a long time ago but I think there was an implication on his part that owning a gun wouldn't be such a bad idea. We are surrounded by hunting camps all of which are populated with heavily armed men the first two weeks of November during deer hunting season, so we certainly wouldn't be out of place.

This advice was filed away and in the meantime Morgan the Wonder Dog came to live at Sunflower Farm. Morgan had a dramatic effect in making the place seem more secure. Morgan is a Shetland sheepdog, a "Sheltie," but I don't think he's a purebred because he's much bigger than other Shelties that we have seen. While he's not huge like a German Shepherd he can still seem very intimidating to some people. We've had a number of people, including couriers, who won't get out of their vehicles until we come and get him to stop barking. It's one of the things that Michelle and I have appreciated the most about Morgan, that he barks at vehicles when they pull into the driveway. This is especially handy when we're in the house and don't hear the vehicle pull in. Most of the time it's someone we're expecting, but sometimes it's a strange car or truck.

I used to find it disconcerting to be coming in from the garden and suddenly seeing a stranger standing in front of me. I often "zone out" when I'm gardening or splitting firewood, lost in deep thought. Or if there's wind in the poplar trees it can mask the sound of a car pulling in. With Morgan there is no mistaking the fact that a vehicle has entered the driveway. He'll even bark if a vehicle slows down or stops out on the road. We've noticed a weird behavior exhibited by drivers on our road. We've noticed then when a person needs to stop on a quiet, lightly populated road like ours, they will wait to do so near a human establishment. We are in the middle of a 10 km stretch of uninhabited road. You can drive 4 kms in one direction and 6 kms in the other direction and not see another establishment. And when people need to stop for some reason, they seem to be most comfortable doing so in sight of human habitation. They might need to stop to change drivers, to check a map or their GPS, to dump an ashtray, to have a pee, to stretch, to get something out of the trunk or to check a tire that's losing air - the list is apparently endless because several times a week we'll hear a car stop and idle out by the driveway and then drive off a short time later. They could stop anywhere in that 10-kilometer range, but apparently to stop in a section of the road where there are just ponds and forests invites attack from the zombies hiding in the woods. Better to stop near a home. While the woods continue to grow in and restrict the view to our place from the road, it's easy to see my 100-foot tower and wind turbine. I'm surprised that anyone thinking of stopping near my place wouldn't consider the likelihood of a misanthropic, crazy, gun-toting hermit of a man living under that tower who might not welcome their stopping in front of his home.

But luckily Morgan makes it very clear that the house is inhabited and should they decide to come down the driveway to 1) ask for directions, 2) ask to borrow some gas, or 3) ask for a tour of our off-grid system, etc. they will encounter a slightly intimidating and loudly barking dog. I particularly like having Morgan when I have to be away overnight when I do workshops throughout the province. On the odd occasion in the winter when Morgan has been inside when a stranger knocks on the door, he makes one heck of a racket and if I were on the other side of the door I wouldn't be anxious to go in. I think for most rural dwellers a loud dog is the best system of home security they'll ever need.

There are some exceptions though. One winter night after dark a car pulled into the driveway. The headlights of the vehicle lit up the living room. I went out on the porch to see who it was. The driver got out of the car, leaving it running, and told me that a car was on fire just down the road from us. He said he was on his way to Ottawa and his cell phone didn't have service and so he asked if I'd call the fire department. I agreed to look after it. I called the fire department first, and then I got in my truck and drove down the road to investigate. I found a truck on fire about a mile down the road (1.6 kilometers.). It was down an abandoned driveway, well off the road. The truck was really burning. In fact, the cab was so engulfed in flames that it was readily apparent to me that if someone was in that cab, there was nothing I could do for them at this point. I have to admit; I have had very little experience with burning vehicles - actually none. So I found the episode quite disturbing. As was so often the case in our move to the woods, this was outside my comfort zone.

I went home and then returned to the truck fire once the local volunteer firefighters had arrived and they had put the fire out. The next morning an O.P.P. officer knocked on my door to ask about the previous night. First he asked if I'd written down the license plate on the car that had pulled into our driveway. I explained that the vehicle's headlights had blinded me and I couldn't even see the driver, let alone the license plate. Then he asked if he could see the boots that I had been wearing the previous night. I suddenly realized that I was apparently now a suspect. I guess it's a cop's job to investigate everyone. I said, "Sure, I'll show you the boots that I had on last night, but I can assure you that none of the footprints around the truck were mine." I didn't think seeing my boots were going to help since I'd bought them at Canadian Tire and 90% of Canadian men own the same boots. And I came clean. I told him that when I saw the burning truck, well engulfed in flames it was way outside of my comfort zone, so I just drove home and waited for the fire department to arrive. He seemed pretty content with that. I guess I can pass myself off as being a wimp better than I thought.

Later I learned that the truck had been reported stolen and torched apparently for insurance purposes. But there was enough weirdness that evening to convince me that situations can arise when you're on your own that require a little more independence that you might expect in the city. In the suburbs we were 3 minutes from the local police headquarters. While I feel very safe where I am now, I also feel I may find myself in a situation where I might need some personal protection.

And so began my journey to own a gun.

Canadians have a different relationship with guns than Americans. Oh we have them, but we don't necessarily flaunt them. We make it very difficult to own a handgun and we have a "long gun" registry, which requires anyone who owns a rifle or shotgun to register the weapon with the government. Most Canadians live in cities and they are pretty happy with the Long Gun Registry. They don't own long guns so it doesn't affect them. And they like the fact that it's difficult to own a handgun. Rural Canadians, on the other hand, who hunt and farm and compete with animals that eat the crops they're trying to grow, are not as thrilled with the Long Gun Registry.

To purchase a gun in Canada you need a "Firearms Acquisition Certificate" or FAC. So at some point after the "truck on fire in the woods" incident, I decided it was time to get mine. This was not an easy decision. Guns kind of freaked me out. I'd owned a BB gun when I was a kid and I shot my Dad's '22 when I was a teenager but I hadn't really been around guns for 25 years. The closest I'd been to a gun as an adult was when I was in line behind a cop at a Tim Horton's donut shop. I always thought it was kind of weird that they carried around this piece of steel strapped to their waist that could kill someone with the pull of a trigger. I'm sure they probably get the same feeling at times too.

So I got the books and did my reading and passed my test. Then it was time to get a gun. I got my neighbor Ken involved because he worked at a penitentiary and he is a hunter, so he knows guns. One of Ken's friends had been in charge of the weapons at the maximum-security penitentiary where they both worked. He also owned a small gun shop in our town.. So off we went to buy a gun. What a strange concept it was for me, the guy from suburbia.

At the shop we discussed various guns but decided on a shotgun. There was a 12 gauge with a large choke on the end. It was quite an intimidating weapon, which was my biggest goal of the project. The choke looks like a silencer you see on handguns in the movies, but is used to control the pattern of the pellets out of the end of the barrel. With a shotgun you can fire slugs, one piece of metal like a large bullet, or you can buy a shotgun shell with pellets. These are round balls of lead or steel that spray out of the barrel when fired. You can either choke them to a very tight pattern; say the size of a Frisbee, or a much wider pattern, like a hula-hoop. The advantage of using a shell like this is that you don't have to be accurate. As long as you are close to a target, there's a good chance some of the pellets are going to hit it. And that's the beauty of a shotgun. You don't necessarily have to be accurate. Just in the ballpark is often good enough.

When I do my "sustainable independence" workshops I'm usually at a college in an urban area. When I get to the section on security I show a photo of a shotgun. The photo is usually met with an uncomfortable silence because I think guns freak most urbanites out. First, I have to explain my situation and that I've been told, by those responsible for upholding the rule of law in my area, that I'm basically on my own for a while after I call them, unless I get lucky and one is having lunch in town and can get to me faster than usual

Second, I explain that the advantage of a pump-action shotgun is the sound it makes as you move a shell from the magazine into the chamber. You can load three to five shells into the magazine, waiting to be used (3 in Canada, 5 in the U.S.). When you're ready to shoot you pull on the wooden slide or fore-end and it takes one shell into the chamber ready to be fired. Doing that makes the very distinctive "che che" used-shell-out, new-shell-in sound. You know the sound; you've heard it in movies. Arnold is constantly doing it in *The Terminator* movies. It is a sound that intimidates. It says, "The person at the top of the stairs has something that will do me a lot of harm if I choose to go up them." And with a shotgun, after that first shot you can quickly pull the next shell into the chamber ready to fire again. The person at the bottom of the stairs basically has to ask themselves, "Do I feel lucky today?"

When you make that distinctive cocking sound with the shotgun, it's virtually impossible to tell if a shell has indeed been loaded, so even if you don't want to fire a weapon, you can still cock the shotgun and point it at an intruder and that intruder still has to ask the same question, "Do I want to risk the consequences of them pulling the trigger, or do I want to leave quickly?"

The person that's in your house will be well familiar with the sound of a shotgun, and even more aware of the consequences of it being fired in their direction. They are going to be sprayed with pellets and if it doesn't kill them instantly they are in for a ton of sorrow for their efforts. I would hope that anyone hearing that sound would choose the more logical option, which is to leave quickly.

And so at the gun shop that night I bought the shotgun. The shop owner said that he would register it for me over the internet and it would be available for pick up in a couple of days. When I went back to pick it up, he hadn't been able to register it. I can't remember if he had forgotten or had had trouble with the website which was new and probably problematic, but regardless I was taking possession of an unregistered weapon. This was during the time when the Federal Long Gun Registry was just being set up, so there was much confusion about the process. The government was being challenged to get long gun owners, who were mostly extremely law-abiding citizens who choose to own a gun to hunt, to register them. So there were still lots of unregistered long guns out there, but the difference was that I now had one. In fact, now I was driving home with one in the trunk. It's amazing how you can go from being a pretty mellow, normal sort of ex-suburbanite one minute, to feeling like a felon the next.

I thought, "What if I got stopped by a cop on the way home? What would I do then?" I told myself to relax. I hadn't seen a cop on my road in years and in fact, at that time of the night I was unlikely to even pass another car in the 10-minute drive home.

So I managed to get home safely and I carried the gun into the house. I placed it on the dining room table. Michelle and I looked at it with a mixture of reverence and horror. What the heck were we doing with this miniature weapon of mass destruction? How could non-hunting pacifists suddenly be in possession of such a symbol of violence and mayhem?

But the moment didn't last long. Suddenly there was a knock at the door. When I went to the door I was shocked to discover a police officer standing on my front porch. It was winter and Morgan had been sound asleep inside the house, which is why he hadn't warned us about our visitor. At the sound of the knock he began barking, and I used this as an excuse to head outside to talk to the officer on our front porch. Obviously the last thing I wanted to do was invite him into our house while a newly purchased, unregistered shotgun was laying on my dining room table. I mean really, what person in their right mind keeps a shotgun on the dining table anyway? I cannot remember the reason for his visit, but as was often the case it was probably about a car that had gone off the road or some other vehicle-related incident. After a few minutes out in the cold I decided it was silly to be worried about him seeing my new gun. I was a licensed gun owner. I had a "Firearms Acquisition Certificate." Heck, the reason I had just bought a shotgun was because one of his buddies had suggested that I should. The reason it wasn't registered was because I hadn't had a chance to do it. I had nothing to hide so I invited him in.

To my relief, my amazing wife had tactfully moved the shotgun from the dining room table to a much more subtle location in the hall closet.

This has been my relationship with guns ever since that evening. I somehow feel like I'm doing something illegal just owning one. And shooting it still terrifies me. I still think I need to buy a 20-gauge shotgun because using a 12-gauge shotgun is like holding a small piece of field artillery when you fire it. It just about knocks me off my feet. I have to brace myself as if I'm standing in a hurricane-force wind to stay balanced.

The one thing I've learned about guns from a security standpoint is that if you have it for protection, you need to use it a lot and be comfortable with it. There's no use fumbling with it when you need it. You have to use it often enough so that you're going to be able to operate it as desired and hit the target. I continue to believe this is the advantage of a shotgun. You don't have to be completely accurate. You just have to be in the ballpark. And you can even fire a shotgun from your hip and be in the game if you're not comfortable firing it from your shoulder.

I have never had to use my shotgun in self-defense or for any other purposes. Last spring we had a mother black bear in the backyard. She was there at dusk with her small cub. They were near the raspberry patch which didn't have any fruit yet. They actually just sat down and ate clover. It was a beautiful thing to watch. Amazing. Exhilarating. The thought of using a gun on one of these beautiful creatures is very foreign to me. Michelle and I watched from the back porch for a long time. Eventually the bears wandered off into the woods behind the horse barn. At the time we had several of Alyce's cows there, Dexters, which are black in color. As the bears rambled by, both of the cows came over to the paddock fence and watched. And as the bears plowed through the bush around the paddock, the cows followed them intently. I'm sure it was a novelty for the cows to have two large creatures on the outside of the fence, but we were convinced that the black cows looked at these black bears and decided that they were related somehow. They were all about the same size and all really black. It was absolutely fascinating. I think if I had wanted the bears to leave, all I would have had to do is make some noise and they would have raced off quickly.

While I do not relish having to use a shotgun, I do have one, and would use it if the circumstances warranted it. I have lived my life in a time of plenty. I have been fortunate to grow up in a time when the difference between the well off and the less well off hasn't been that great. But that gap is widening and as the great recession wears on, more and more people are feeling they are being left behind. In Canada we still have lots of social programs to make sure that people are not destitute. But we also have record levels of government debt, and an aging population set to bankrupt our universal healthcare system at a time when governments are going to be forced to deal with the massive challenges of peak oil, resource depletion and climate change. I am afraid that the number of people that will fall through the cracks will rise and some of these people will become desperate and desperate hungry people often do unpredictable things. I am always happy to help someone in distress but I have spent the last 15 years of my life learning how to make myself independent. I have prepared for a rainy day. I have learned to grow and store food. If someone thinks that they can come and take that from me because they chose to spend their time in other pursuits, they are sadly mistaken. I will not go down without a fight.

If I sound like a redneck then so be it. As a long time lefty I have been fully supportive of socialistic policies like universal healthcare and income support. But some people have grown up expecting the government to look after them from the cradle to the grave and I believe that is an unfortunate, unintended consequence of social policy. If someone is in dire straits and wants to approach me about working here for room and board, that is something I will consider at the time depending on the situation. If someone wants to come and take the food that I've grown and stored, well, to paraphrase the NRA, they'll have to pry that bag of potatoes from my cold dead hands. Or there may be cold dead hands at the end of the negotiation, but I don't intend them to be mine.

In the future I think there will be more income disparity and more unrest. I think it will be most common in urban areas, where the bulk of humans choose to live. I also believe that cities lack that feeling of community that comes from living in a rural location. In tough times small communities pull together. There is a sense of common purpose. Of shared goals. When something bad happens to a member of a small community, it's like a member of the family has fallen on tough times. And good things happen as a result. I am always amazed when a family loses a home to fire or some other catastrophe, how quickly their immediate needs are met through local contributions, and then how much of the materials and labour needed to rebuild are donated.

Belonging to a community is the best security you could ever have in uncertain times. When you live far out from the local town or village, a big, loud dog is the next best defense to avoid bad situations. Good lighting and even an alarm are also excellent deterrents. And finally, owning a gun means you've crossed the Rubicon of self-defense and are ready to take things into your own hands if need be. While I hope it never comes to this, I take some comfort in knowing that the option exists at my house.

18 Show Me The Money

by Cam

"Things are thieves of time. The more things you have the more you have to work to get those things. Frugality is the wise use of resources."[1]
Nathan Gardels, editor of the *New Perspectives Quarterly*

When I was a kid I had a Globe and Mail newspaper route. It was a morning paper and we lived in a subdivision on the St. Lawrence River. The subdivision was populated by businessmen and university professors who loved the Globe, but I didn't understand it at the time. I had to get up at 5 a.m. and would deliver my papers by 6, which gave my subscribers time to catch up on the news before they left for work. Years later when I began to read the Globe myself I finally understood why they gave me five dollar tips at Christmas. It was because they really liked getting the Globe. If I had known it at the time I would have doubled the rates, and I'm sure they would have paid them. Five dollars seemed like an outrageous amount of money for a 12 year old in 1971 and I believe I got the odd $10 tip as well. How I wish I'd clued into the leverage I had at the time.

I saved my money. I used to cut lawns and baby-sit and do just about anything I could to make money. And I saved it. I lived out in the country and didn't really have anything to spend it on. Everything was taken care of by my parents. So I'm not sure where this desire to save money came from. And I started a vegetable garden when I was sixteen. If I were reading this book I'd probably say "What a strange young man."

I believe it was because in a former life I was a poor farmer, so I was predisposed to this in my DNA. Regardless of why, I can remember having $1,000 saved in my Royal Bank account at a young age and my parents constantly needing to borrow it. It used to drive me crazy. "But we'll pay you the same interest" was always their line. Which they did. But it was just way better when I could take my bankbook in and get it updated and the bank paid me interest. Free money for not doing anything. Printed in my bankbook!

I continued to have jobs during high school and save money but not proportionate to those early days. As I got older there were more things to spend it on. Albums (yes, vinyl albums!) Gas for dates. Burgers. Michelle and I were living together eventually after high school and there was rent, and money for a car. And school. University. Eventually Michelle was teaching and I was selling computers. There was an investment counselor in the building where I worked who had discussed computers with me, so I made an appointment with him to invest in the stock market. He went through an inventory of our assets (none, we rented an apartment and had a car loan) and our incomes, which were moderate at that time, compared to many, but seemed pretty good to us. Then he asked how much we had in savings. "Savings?" I asked. He said, "You make that much money and have no savings? Get out of my office and don't come back until you do!"

I think I was about 25 at the time but felt as though I were 7 and my father had just yelled at me for setting fire to my plastic soldiers in the sandbox (not that I ever did that). I was mortified. From that day on Michelle and I started saving money, and we've been savers ever since. That's the advantage of getting advice from an outside person. Often you'll do a better job of getting your house in order for them than you will for yourself. I'm sure there are many books written about what in our makeup causes humans to behave like this. I'm guessing it goes back to pleasing our parents, but that is another book. And who cares? Once I started saving it was like: "Ken (my financial advisor) will be so proud of me!" And sure enough six months later we went back and he was impressed and decided to take us on as clients.

Eventually we were able to buy a house and our daughters Nicole and Katie came along.

About 20 years ago we needed to renegotiate our mortgage. Our existing bank showed us what the new monthly payments would be. A second bank did the same. Then I went to meet with someone at Canada Trust (now TD Canada Trust.) The representative showed us what the monthly payments would be. Then they asked if we'd be interested in paying weekly. Our monthly payments were going to be about $800. I suggested that we did not have the financial wherewithal to pay $800 each week. She replied that the net monthly amount would be the same, $800 per month, but that $200 payments would be taken out of our account weekly. Then she swung her computer screen around and showed us that if we paid weekly we would save approximately $35,000 over the 20-year life of our $66,000 mortgage.

I was blown away! First at how much money I would save, and second at how even though my current bank offered weekly mortgage payments, they hadn't mentioned them to me. Why would they if they would lose a huge chunk of profit? They really didn't have my best interests at heart, but it is a capitalist system after all.

I asked the customer rep why everyone wouldn't pay his or her mortgage this way? I assumed that most people weren't aware that this was an option, but she said that most people's finances didn't allow for it. They were paid monthly or biweekly and their budgeting didn't allow weekly withdrawals for the mortgage. This is bad financial planning on a number of levels. Mostly it's bad because people live so close to the edge in terms of money in their accounts. This is the wrong way to manage your money. You need to have a slush fund amount in your account that allows you to take advantage of opportunities like this.

Needless to say we switched our mortgage to Canada Trust and we also looked at some savings we had. The Canadian government at the time gave checks to parents called a "Baby Bonus," which was designed to ensure that children had food and clothes and basic necessities. Under the concept of "universality," every Canadian child received these payments, regardless of the parents' income. Because we were frugal we were able to bank these checks. We called the account the "Kids' "College Fund" account. Eventually it grew to about $10,000.

This was about the time we were starting to feel a desire to move out of the suburbs, and we knew the key to our financial independence would be paying off our mortgage. So we took $9,900 and put it towards the principal of our mortgage. Our new mortgage had a feature where once a year on the anniversary date of the mortgage we could pay off up to 15% of the principal. So that $9,900 reduced our principal from about $65,000 to $55,000. That was a pretty great feeling, but it was tempered by the realization that this would probably be the only time we could do this.

As the year went on though, we decided to give it a shot again. We took a percentage of every one of our paychecks and put it into our "Five-Year Plan — Pay Off the Mortgage" account. This had previously been the "College Fund." We scrimped and we saved and we put off buying things. We wanted to take the kids to Disneyworld, but that was $3,000 we could put towards the mortgage. We needed a new bed, but flipping the mattress over would have to do for another year. The rust had eaten away at the floor under the driver's side of the car to the point where you could see the road whisking by, sort of like a Flintstone-mobile, but a new car could wait.

Low and behold, the next year we were able to do it again, and we put

$9,900 towards the principal on the anniversary date of the mortgage. I should point out that we were by no means well off in terms of income. The median family income at the time was about $50,000, meaning half the families in the country had incomes below that figure and half had incomes above that figure. We were always comfortably in the lower half. In fact, as a result of the challenges of running our own small business there were some years when our income dropped dangerously close to what in Canada was deemed the poverty line. Canada has a very generous definition of what this income level is, but I share this to show you that you don't need a high income to become financially independent. You need to be frugal and you need to get out and stay out of debt. But you have to be solely focused on this one goal. A weekly trip to the mall will not help you in this cause.

Each year as the car got older and older and we wanted a new one, we held off. Eventually whenever we took a long trip we'd rent a car. Owning and operating a car costs many thousands of dollars a year in taxes, insurance, maintenance, and repairs. We took collision insurance off the car because it was so old that if we had been involved in an accident they wouldn't have given us much anyway. But it was paid off and we elected to save a huge amount of money and keep driving that older car and put the savings towards the mortgage. Keeping the beater on the road and renting a car for longer trips that required a more dependable car saved us thousands of dollars.

For three more years we continued in our single-minded mission of paying off our mortgage. Snowsuits got used a third year when they were probably tighter than the kids would have liked, we came out at the winning end of our Automobile Club membership as we made greater use of tow trucks and services than someone with a newer vehicle, and vacations were canoe trips in provincial parks.

The other brilliant part of paying off a chunk of principal on your mortgage periodically is that as you continue with your regularly monthly or weekly payments a greater percentage is going towards the principal. Mortgages are front-end loaded, so that in the early years of the mortgage you are paying mostly interest and in later years you are paying mostly principal. So what you are doing with these regular principal payments is moving up the time continuum so that a larger percentage of each payment goes towards the principal.

Let's say you had a mortgage of $240,000 with a term of 30 years and an interest rate on the loan of 6%. The national average is that people sell their home every 7 years. If you did this you would owe about $216,000 on the principal after having paid $120,000 in mortgage payments. Of that amount only $24,000 would have gone towards the principal. You would have paid $96,000 in interest.

On the 5th anniversary of our mortgage we were able to apply one last check and pay off our mortgage. Of all the feelings in the world, there are few that rival the feeling of leaving your bank without a mortgage. We photocopied our mortgage and had a ceremonial burning in the fireplace that night. (We thought it was a good idea to hang onto the original just in case.)

It was as though a huge weight had been lifted from our shoulders or the storm clouds had left our home and the sun had come out. Suddenly anything was possible.

And I think everyone should experience this same feeling of freedom. It's going to require sacrifice though. It's going to mean keeping that downhill ski equipment a couple of years longer than you'd like. Heck, it means not downhill skiing at all. It means heading to the local ski swap or reuse centre, picking up a pair of used cross country skis, and finding a forest or trail near your home to ski on for free. It's cheaper and better cardiovascular exercise.

Paying off our mortgage was probably the key to us being able to move to the country. Once we got that monkey off our back it made anything seem possible. It meant if we had a bad month for business, the bank wasn't going to come and take our home. And it let us start putting money aside into various savings accounts. These were "5 Year Plan Accounts." Once we decided it was time to move out of suburbia and find a place far from the maddening crowd, we just took all the money we had been saving to put towards the mortgage and put it into savings accounts. This would allow us to have more flexibility in the years ahead.

The other thing in our favor that allowed us to move away from the city was the fact that we were self-employed. We ran our business and while most of our customers were in the same city, they were becoming used to not seeing us that often. We did electronic publishing, and they were used to receiving proofs in PDF format on their computers. Traffic everywhere is getting insane, and even a short drive around suburbia can waste a ton of time, so we just used technology to reduce how much time we spent on the road. Then when we finally did move away, customers were already used to not seeing us that often. I used to say we could be working out of a french-fry truck and no one would have noticed. Other than the fact that our artwork might arrive covered in grease and I'd have gained 100 pounds in a couple of months.

With technology I think working from a distance becomes more and more feasible for more people. It just worked out that we were able to refine our business to allow it to be possible for us. The downside to being self-employed is that we were subject to the whims of the market. So Michelle and I always made sure we had financial cushions. We kept extra cash in the business and we kept money in savings accounts personally. This is not how most North Americans live their lives. The average family spends the money they make, and before the economic collapse of 2008, most spent much more than their income. They were able to do this by using equity in their homes to get lines of credit, which they could use to buy more "stuff." This always seemed bizarre to me, but it was the norm in those heady days of the housing boom. If Michelle and I didn't have the money sitting around, we didn't spend it. And after we had the mortgage paid off, and we did start having money sitting around in a savings account, we left it there. It was our "rainy day fund" and buying a new vehicle didn't qualify as a rainy day.

I'll reiterate that Michelle and I also never made that much money. I do this because it's easy for people who have their own dream of a getting away to feel discouraged, that this is something for wealthy people to do. But that was never the case with us. It was always based on spending less money than we earned, and always putting money away into savings.

We got into the "pay yourself first" mode. When we took a paycheck out of the business some of it went into savings. Every check. Every month. This way the money was not available to be spent. When we were saving to pay off the mortgage this became a holy crusade. And we enjoyed it. I think people like having a goal like this and it was our all-consuming passion.

We were very grateful to have been able to afford a home and we are very aware of the fact that timing is everything. David Foote, the demographer who wrote *"Boom, Bust and Echo,"* seems to suggest that my parents' generation had it pretty easy. The North American economy was growing rapidly and after World War II there were always jobs. Unless you were a real screw up, you basically got dragged along by a society that was generating wealth exponentially and you were getting your share. When I graduated from high school in the late 1970s there were still some people who got half decent jobs without post secondary education. Unions were still strong enough to help lots of middle class people have a good standard of living. When we bought our house in 1985 we paid $80,000 for it. While we weren't making buckets of money I'm most aware that this was still very affordable.

Now I'm not part of the demographic group that bought their house in Toronto for $10,000 in 1960 and sold it in 2000 for a million dollars, but I am grateful that I was able to purchase a 2-bedroom house for $80,000. And that it was worth $155,000 when we sold it 11 years later. I am not one of those "well look at me, I'm a self-made guy" individuals. I got lucky. I was born in the right place at the right time.

What I will take credit for was deciding not to take 25 years to pay that mortgage off. I've got to say that after living off grid for 13 years, I'm not sure I would have taken on this challenge when I was 50 or 60. I was 39 years old when we made the move. And with the work we do here, and heating with wood and cutting all my own firewood, and having a huge garden, all while running a business, I'm not sure I'd have the stamina if I'd started in my 50s. Not on the scale we've done it. It's been a ton of work, lots of it physical, and it's helped that I had the energy.

I suppose if I were older and had saved more money I could have bought a tractor, and paid someone else to cut my firewood, and grow my vegetables. But I'd have missed out on the best part. And I've been able to eat whatever I wanted all that time because I know I'm going to burn those calories off.

Earning an income since moving to the woods has been much more difficult. It has been hard to develop new customers. This is partially because the nearest city to where we live, Kingston, is smaller than Burlington where we started the business. Burlington is a suburb of Toronto and the Greater Toronto Area (GTA) has more than 5 million people in it, versus 120,000 in Kingston. We did maintain some of our customers in Burlington for years, but eventually it became too difficult. Mostly because at some point they still wanted to see me, and that meant a 3-hour drive to the far side of the GTA. This meant that I got to drive through the entire length of some of the most congested highways in North America.

I dreaded these drives. I was depressed the night before I went. I was stressed out the entire day I was gone. And I was bagged the day after I got back from the stress of the drive. And from an environmental point of view I was in the car most of the day. Yes, it was a small Honda Civic with great gas mileage, but it was a car. At a certain point, it was just getting too hard to rationalize living off the electricity grid to reduce our footprint on the planet, then spending a day on the road every 6 to 8 weeks.

In 2003 my uncle Ian Micklethwaite approached us about getting involved with some workshops he was doing on renewable energy. He published a magazine called *"The Farmers Finder"* and it was delivered free to rural mailboxes across the province. So we ended up doing 3-hour workshops for 4,000 people in 7 cities throughout Ontario. From this Ian started a magazine called *"Private Power, For People Who Want to Produce Their Own Electricity."* Michelle and I edited it. Well actually, we wrote many of the articles, solicited articles from others, edited the magazine, laid it out, sold some ads, and provided the final artwork to Ian's printer. It came at a busy time for other things we had on the go, but we were able to manage.

One of the writers I was able to get an article from was William "Bill" Kemp. Bill also lived off grid and he wrote a Primer for Off-Grid Living. Bill had some spare time and while he was an engineer who builds hydro-electric stations and biogas systems, he seemed to quite enjoy the writing. When he was done he suggested next he would write an article about inverters, then maybe one about solar panels and one about wind power.

So one day I met him for coffee and I had mocked up a cover of a book called *"The Renewable Energy Handbook for Homeowners"* by William Kemp. And he liked the idea. So Bill wrote an exceptional book and we began learning about the book publishing industry. It turned out our timing was very good. As we had discovered when we moved off grid there was not a good resource available anywhere. There was an out of date book on solar power, and a technical book on wind power, but nothing that put it all together. Bill started with energy efficiency, and then worked his way through solar power, wind power, batteries, inverters, and generators and tying it all together. And it really struck a chord with readers.

So more and more of our time was taken up with the book. We started with a Canadian distributor that did a lousy job of selling the book and then went out of business owing us $20,000. Then we got another distributor in Canada who had competing books and while the books sold well on their own, and we were getting paid, they were taking a large percentage of the sales and promoting their own books ahead of ours. So when our contract was up we went out on our own. By now Bill had written a book called *"Biodiesel, Basics and Beyond"* and *"The Zero-Carbon Car,"* about building a plug-in electric hybrid.

As sales became more dependable with the books we progressively gave up customers in Burlington. It was kind of a dream come true.

We were actually able to earn our income from pursuing our passion for renewable energy. We purchased a good quality video camera and began making educational DVDs as well as our books. One summer our daughter Katie was home from university and we paid her to film me all summer in the garden. The DVD took the viewer through every aspect of growing vegetables. Katie shot and edited it together using iMovie. Our *"Growing Great Veggies"* DVD continues to sell well.

After a few good years of book sales being consistent we finally gave up our last and biggest customer from the city. The timing was good because I was pretty burned on laying out catalogs for industrial equipment. So things really seemed to be going well.

Then the bottom fell out. The economic collapse of 2008 had a dramatic impact on our business. The majority of our sales were to the U.S. and when the U.S. housing market collapsed, demand for books went with it. Turns out when you're underwater on your mortgage, buying books about sustainable living end up low on your priority list, so we saw a dramatic drop in sales, and in our income.

Finally we were practicing what we preached. I had suggested in my book *"Thriving During Challenging Times"* that living a frugal life made sense on many levels. We had no debt. We had no mortgage. We did not require much money to get by. We lived humbly. We had no electricity bill. We had no heating bill. We were using progressively less propane each year as we invested in more solar panels and a solar domestic hot water heater. We had a pantry full of rice and pasta and canned goods, enough to last us 6 months, and while things weren't so bad we had to use it, it was there if we needed it. And we had a rainy day fund. For that matter Michelle and I both had some money in our retirement funds. While it wasn't much, when you hear how many North Americans have no retirement savings, it helped reduce our stress just that much more.

And from adversity comes creativity. We've been busy reinventing ourselves over the last several years. And we've started charging for things we previously did for free. We have always found lots of people interested in visiting our off-grid home. Now we offer a three hour tour and sit down discussion of living off grid for $150. And a surprising number of people are happy to pay it to hear what we've learned over the years.

And the guesthouse, which was always excellent when friends visited, seemed like a natural income generator. So we now have a Bed & Breakfast and Renewable Energy Retreat. People can come and stay overnight like a traditional B&B. Or they can come for a weekend and spend the whole time seeing what's involved with living off the grid.

I think people get excellent value when they come. Michelle and I have spent 14 years learning everything we can about sustainable and independent living. We can help guide people through the enormous learning curve and avoid making the mistakes we did.

For the last 15 years we've been mortgage-free and that allowed us to uproot our family from suburbia where we had an electronic publishing business and move it three hours away to the woods. It also allowed us to evolve the business from doing work for corporate customers to publishing our own books about renewable energy and sustainability. It's unlikely we would have had the confidence to undertake such a journey if we hadn't been free from the constraints of a mortgage.

We also made a decision to only purchase a rural property we could pay cash for. We were not going to take on another mortgage after paying our old one off. It's too enabling a feeling to be free of the obligations of a mortgage to ever go back to those days.

I'm not a big fan of debt. I fear it. And I don't want to go there again if I can help it. If I can't afford something I'm not going to purchase it. That attitude has allowed us to move where we wanted to live and pursue our passion to spread the word about sustainability. While I sometimes regret the loss of income from our corporate days, I do not regret selling my soul for a good income. Ultimately I think, at the end of our lives, we'll have to reconcile the way we've chosen to live. I believe that the evolution Michelle and I have taken on our journey has been logical and rewarding. I am very grateful we had the kind of foresight we had 25 years ago that set us on a path of financial independence and allowed us to pursue this path.

And every day as I sit at this computer and I stare out at my solar panels and my wind turbine and my pond and the forest that surrounds us, I am grateful. Life is good.

19 Trouble on the Horizon

by Cam

We moved to the woods because I was not cut out for city life. I felt out of place. I didn't like shopping. I didn't like cars. I didn't like noise. I didn't feel like I wanted to invest time in my small city lot, because it didn't have the feeling of permanence. I longed for open spaces without human created structures always in view. I wanted that feeling I got on canoe trips at night, when I lounged on a rock and stared up at the night sky and viewed 360 degrees of infinity and gazed in wonderment at just how insignificant problems can seem when you can see the big picture.

After spending many years advocating for the environment in our suburban environment, I was also very tired. I was exhausted from trying to change the paradigm from a car-centric, consumption-based framework that never factored in the impact of planning decisions on the planet. I had been involved with a small group of citizens that had fought a huge infrastructure project. The Town of Milton, which was in the northern part of our region, had reached its maximum potential for growth because the aquifer from where it drew its water could not support any more development. Land speculators, hoping to profit from the constant march of subdivisions that was the norm in the Region of Halton where it was located, had purchased much of the farmland around Milton.

So the developers started demanding more water to build more houses. The only way it could be provided was to build a huge pipeline from Lake Ontario in the southern part of Halton, to Milton. This was going to cost $500 million dollars, and those profiting from new growth in Milton were not going to pay for it. The existing tax base in Burlington and Oakville would pay it for. So we lobbied City Hall and Regional Council, and held public meetings to try and let citizens know their taxes were about to go up to help put more houses in Milton, even though there was lots of infrastructure that already existed that needed upgrading. We got lots of good press coverage, but the pipeline was built anyway. As I drive past Milton now it has grown prodigiously, with more tract housing and more big box stores, looking like every other suburb in North American. In fact Milton had become the fastest growing city in the Province. Nope, it could not stay a livable sized "town" - it had to grow.

In working to stop the pipeline, I met an economist named Tom Muir who was key to helping me understand a lot of development issues. Certainly it was clear that the existing tax base was going to support new growth. I had always assumed growth was a good thing, but was learning that isn't always the case. Modern economics considers growth optimal. In fact there is an abundance of information that shows "constant growth" masks many deeper, underlying problems with the economic system. "Cancer is growth" was Tom's mantra. Cells growing out of control can kill an organism. Perhaps human's constant growth on the planet was a bit like cancer. It certainly seemed hard to stop it.

While I was living in the suburbs a decade ago, the potential to harm the environment from constant unchecked growth seemed very abstract. Now, that's no longer the case. Movies like Al Gore's *"An Inconvenient Truth"* and Leonardo DiCaprio's *"The 11th Hour,"* are dramatically illustrating the urgency of changing human behavior to prevent an environmental catastrophe.

George Monbiot, an acclaimed British journalist, wrote the book *"Heat: How to Stop the Planet from Burning."* While the Kyoto Protocol set targets for signatory countries to reduce CO_2 emissions to 15% below 1990 levels, Monbiot argues that even that target falls way short of what realistically needs to be done. He basically says that in order to stop the planet from entering into a myriad of positive feedback loops that will make the planet uninhabitable in short order, developed countries like Canada, England and The United States, have to reduce their CO_2 output by 90%! It is a staggering number, almost unfathomable to many North Americans. But Monbiot provides a very convincing argument for the urgency of the action we must take. He also spends much of the book providing analysis for exactly how we could achieve such aggressive targets. It will not be easy. Our lifestyles will change. We will have to stop flying. But if we are serious about leaving a planet that is a tolerable place to live for future generations, we must take this action.

Monbiot lays it on the line, *"Climate change is perhaps the gravest calamity our species has ever encountered. Its impact could dwarf that of any war, any plague, any famine we have confronted so far. It could make genocide and ethnic cleansing look like sideshows at the circus of human suffering."*

Books like *"The Weather Makers, The History and Future Impact of Climate Change"* by Tim Flannery made it even more clear. Flannery is an acclaimed biologist and has taken a scientist's approach to critically looking at the mountains of data on climate change, and he paints a very stark picture.

Flannery, like Monbiot makes it very clear that we must act now, and act aggressively. He also explains how we have already impacted many of the earth's systems, and there will be consequences, in short order. Many areas of the planet are already facing severe droughts. One of the predictions with climate change is that rainfall will become erratic, raining heavily in one area, and not raining at all in others. These inconsistencies will start impacting North Americans with higher food costs. According to the National Farmers Union, 2007 was one of the worst years on record for grain production. "The world is now eating more food than farmers grow, pushing global grain stocks to their lowest level in 30 years."

Climate change had produced a 10-year drought in Australia, one of the world's largest exporters of grain. Then in 2010 it was hit by historic floods. Other areas are either too wet or too dry, which has reduced crop yields. Meanwhile, farmers have to deal with fertilizer costs that are significantly higher because most of it comes from natural gas, much of which we don't produce in North America anymore.

Huge parts of the world are used to dealing with droughts, but the southern United States are having such problems, it's becoming questionable how long the water supply will hold out in cities like Atlanta.

While climate change is creating huge challenges for the planet, dwindling energy supplies are starting to cause huge problems for the dominant species on the planet. In the 1950s, a geologist for Shell Oil, M. King Hubbert, had a theory that the United States would hit "peak oil" in 1970. Peak didn't mean that U.S. would run out of oil, simply that it would be the time when it produced that absolute maximum amount. After peak, production would go into inexorable decline, and no amount of drilling or wishing would ever alter the decline.

At the time the U.S. was pumping a huge volume of crude oil, which continued to grow every year. The oil establishment mocked Hubbert. In the early 70s, oil pundits crowed about how the U.S. was now pumping 10 million barrels a day and laughed at that "crank" who had predicted back in the 1950s that the U.S. would be at peak by the '70s. But low and behold, as we progressed through the '70s and looked back, 1970 was indeed the year that the U.S. produced the most oil ever. With rapidly increasing prices in the 70s because of OPEC, you had a tremendous amount of activity drilling for more oil in the continental U.S., as well as Alaska and the Gulf of Mexico, but this just masked the fact that the

U.S. had hit peak.

Hubbert also had a theory that the world would be hitting peak oil right about now. There are many indications that he might be correct. Crude oil in November 2007 was at $100/barrel. In the fall of 2001 after 9/11, oil was about $20/barrel. This is a phenomenal jump in a very short period of time. China and India's demand for oil is increasing drastically, and according to the International Energy Agency (IEA) there is a shortfall in the demand for crude oil and the supply. Potentially, the world will never produce more than 85 million barrels of oil a day.

One of Hubbert's co-workers at Shell, Kenneth Deffeyes, in his book *"Beyond Oil: The View from Hubbert's Peak,"* suggested that we need to look at the behavior of oil companies for signs that they believe we are at or close to peak. With the price of oil at an all time high, it would make sense for oil companies to be going on a massive exploration binge, but instead they are hoarding cash and paying large dividends. Could it be that they know all the easy oil has been found? They are not replacing oil tankers as they are retired, which doesn't make sense unless you think there won't be the crude oil to fill them up. There hasn't been a new oil refinery built in North America since the 1970s. Why would you invest $500 million in an asset that you may not be able to get raw material for?

As is so often the case in a maturing industry where there isn't room for growth, you see consolidation, as companies have to merge and acquire competitors to grow. In the last decade or so you've seen Exxon buy Mobil, Chevron buy Gulf and Texaco, BP merge with Amoco and Arco, and Total consolidate with Fina and Elf in Europe. So while there seems to be no shortage of oil industry pundits who think that the world can continue to find new supplies of crude oil indefinitely, there is ample proof that we are nearing peak.

And then in April 2011, Fatih Birol, the Chief Economist for the International Energy Agency (the group that advises government about energy issues,) admitted we hit peak oil in 2006. In 2006 he was calling anyone who suggested peak oil was even a possibility, a "doomsayer". But now he's changed his tune and admits we've hit it. He also suggests that governments should have started planning for it a decade ago. But of course a decade ago his organization denied that peak oil was even a possibility.

Peak oil poses a tremendous challenge for humanity. Liquid hydrocarbons have incredible energy density, which means the potential amount of work they can displace is staggering. We have used crude oil as our "man servant" to do our grunt work for us. And since it's starting to run out, it's going to change our lives profoundly.

The potential computer failures of Y2K got some people thinking about how reliant we are on technology. But the new millennium passed with few reports of major problems and we got on with our lives. And now we hear about peak oil and assume someone will find an alternative. But this time there isn't one. It doesn't exist. It's not in hydrogen, or fuel cells or even in the sun or wind. Oh we'll try and make up the difference and burn more coal, and frack more rock to get at shale gas, but ultimately, the peaking of conventional crude oil is going to change all our lives.

While it's not possible for everyone to live off the electricity grid, I think having that mind set helps. I think knowing how you power and heat your home and what will happen if those outside energy sources stop coming to your house, what the outcome will be, is a good thing. Then having a plan on how to deal with it is really important.

On August 14, 2003, I was in my office at 4 pm listening to the radio when it went off air. I didn't think much about it, just switched to CDs. We had some guests staying with us that day and we finished dinner about 8 pm. During this time the water was flowing and the fridge was keeping our food cool. Our daughters went in to the living room to watch TV. They were expecting to watch one of their favorite shows but instead there were special news reports up and down the dial. The girls came back into the dining room and said "Something seems to be happening in the city." And low and behold since 4 pm, 50 million people in the Northern U.S. and Ontario and Quebec had been without power. It meant traffic chaos and a complete disruption of normal life.

I would suggest that these events will become more common. And I would suggest our response will be the same; unchanged. Here at Sunflower Farm we are unaffected by such events. All the systems in our house continue to work fine when the grid goes down. Our lifestyle doesn't change a bit. Now I always enjoyed blackouts when I was a kid. They were fun. Candles. No TV. And in August most city dwellers made the best of it. Dinner at a patio restaurant. Or ice cream eating parties as the stuff in the freezer started to thaw. But if a major blackout happens in January or February it would be a much different story. Within a day or two, big cities would become refugee camps.

I think it's important that people start making plans for such events. And I'm not just talking about batteries and flashlights and bottled water. I suggest that people should start installing independent energy systems in their homes. The technology exists, all that's lacking is the motivation.

In my book *"Thriving During Challenging Times, The Energy, Food and Financial Independence Handbook"* I take a systematic approach to the challenges we face, and then provide step-by-step suggestions for what you should be doing to prepare. Yes, you'll have to prioritize based on where you live, how you heat your home, where you get your food, your employment situation, etc. But it's time to start doing this. In the almost decade and a half we've been off the electricity grid the technology is infinitely more advanced and cheaper. I urge you to start making yourself more independent.

20 Just Do It!

by Cam

Our 14 years of living off the electricity grid have been terrifying and gratifying and fulfilling and challenging and rewarding and frustrating and absolutely invigorating. There have been times when I've thought enviously of people who get one job and live in one house in one city doing the same thing for their whole life. My envy usually doesn't last for long and it usually only comes when some piece of equipment isn't working, or the garden seems completely overwhelming, or the firewood pile for next year isn't getting filled up as quickly as I'd like.

What I've discovered is that by putting my nose to the grindstone and trying to break problems down into manageable chunks I can usually get through. Figure out what the problem is. Figure out what I need to fix it. Try and locate the equipment or expertise that can help. Do the easy stuff first. Work up to the harder tasks. And finally just commit the time and mental energy to finish the job.

Any talk of "packing it all in" have been "momentary lapses of sanity" in my mind. I know I would be miserable back in the city. I know I would be miserable with a regular job. I know I would be frustrated not being able to try and live as sustainably as I can and share the experience with others. But I won't try and kid you that's its been an easy ride.

I think the story I told of my move here in the introduction tells it all. Standing in the garage of our new home, in January, at 2 a.m., in a blizzard, absolutely exhausted, wondering what the heck I was doing moving my wife and daughters 3 hours away from our family and customers and the safe suburban world we'd grown accustomed to. I thought I was losing my mind. What was I thinking? Really! This had to be the most illogical thing I'd ever done, in a string of many.

That night, as I was moving stuff into the guesthouse out of the moving truck I dropped a framed picture and broke the glass. The picture in the frame had been done by our friend Joe Ollman, a talented writer and illustrator. It shows the four of us in the basement office of our suburban home. The girls are there, Michelle is being mellow and I'm freaking out. After we got settled in to this place I thought about getting the glass

The illustration by Joe Ollman of our previous home office. I dropped the framed print and broke the glass in the middle of a blizzard on moving night, and hung it with cracked glass in my office to remind me of the panic of the move.

replaced but I decided not to. I wanted to leave it cracked to remind me of that night. That was a pivotal point in this journey to this place and where we're at now.

Moving to this remote off-grid home was also a logical progression. It was the next logical step. You sort of ramp up to these things. You start by recycling your cans when there's no municipal system to do it. You start riding your bike wherever you can. You join the local environmental group. You start a small vegetable garden. You raid your neighbor's garbage piles for bags of leaves for your compost heap. You pay off your mortgage to give you some freedom. Then you cross the Rubicon. You plant corn in your front yard in suburbia and realize it's time to go. You don't belong there anymore.

Once you step off that edge it makes the adjustment to rural living a lot easier. Fourteen years ago when the technology of solar and wind and inverters was in its infancy there was nothing to prepare us to go off grid. We were cutting the umbilical cord to civilization and saying that we were ready to go it on our own. And at that time it was a cross between fearlessness and stupidity. Especially considering how little we knew about the whole thing.

After home schooling our daughters for years it was the ultimate test in our philosophy of life-long learning. We were simply going to have to figure this all out and stay with it.

And all these years later the system is working fabulously. And we "get it." We get the whole energy thing, how great it makes our lives, how hard it is to make and how grateful we should be for having access to it. And we "get" the relationship between our diet and our home and our impact on the planet and we know what we have to do to minimize it. We don't fly. We drive as little as possible. We've got an electric bike to get into town. We make our hot water from the sun. We power our home with the sun and wind. We grow an increasing amount of our food. We use an ever-decreasing amount of propane and fossil fuels with the goal of using none. Our home will be carbon neutral.

The path least traveled is usually not the easiest one. There are way more bumps along the way. But those bumps make getting to the destination all the more gratifying. I spend an inordinate amount of time walking around our property pinching myself that I actually live the way I do. I look up at the wind turbine making power for our freezer, or our solar panels powering our computers and high-speed internet, and our solar domestic hot water system making our water hot, and I think, wow, I did this! I'm off the grid but I'm living with all the creature comforts other people have, only I make them myself. Oh, I had some help from the folks who designed and built the technology, but I invested my time and money and put my soul into this system and it works amazingly well.

I am no longer dependent on someone else to provide me with electricity to keep my lights on and my food cold and water flowing from my well. I am no longer dependent on someone else to provide me with natural gas to keep my home warm on the coldest of nights. And increasingly I'm not dependent on a food system to keep me fed. I do all of these things myself. We all have the same choice in these matters. I choose self-reliance over dependence. I choose to be in control myself rather than at someone else's mercy. I choose to be resilient to shocks rather than live in fear of the next shock.

Each year we enjoy a visit from Ken and Madeline Snider, who lived in our house more than 70 years ago. Ken's dad had to leave the farm because his doctor said he was working himself to death. So at the age of 14 Ken was forced to take over running the farm. And it was an enormous amount of work. An insane amount of work. It meant cutting all the wood to heat the house in the winter by hand saw and splitting it by axe.

It meant cutting all the hay that was needed for a number of horses that did much of the heavy work, and cows for milk and meat, and pigs and chickens. Much of the "hay" would have been marsh grass which would not have had much protein for the animals, which would have meant they required even higher volumes. In the summer Ken cut much of the hay in marshy areas. He piled it by hand in huge haystacks. Then in the winter once the marshes had frozen up he went in by horse and sleigh to retrieve it. A backbreaking load every day.

You and I cannot conceive of how much physical effort this was. Basically he might have eaten as many calories as he physically could consume, and still have burned more than he took in. Yes, he used horses for much of the heavy work, but just rigging up a team of horses in those days with the weight of the gear would have done most of us in by 9 a.m. Ken had exactly the opposite issues that most North Americans face today. Consuming too few calories. Most of us today consume too many calories and burn too few.

Ken makes me marvel at the wonder of living off grid as we do. How I can take that magic genie in a bottle, "oil" and use it to displace the work of many men and animals. On a Saturday morning I can cut the

Ken Snider explains to Cam what life was like on the farm 70 years ago, while standing in the garden near the barn foundation.

same amount of wood it would have taken Ken and a buddy two weeks to cut. And now I can use electricity to cut and split even more of it without having to consume oil. Plus I get to burn that wood in a woodstove that is much more efficient than anything Ken would have had 70 years ago. Most of his heat went up the chimney. My modern EPA-certified woodstove burns very cleanly because it is squeezing every BTU of heat energy out of that firewood that it can.

Having had Ken visit regularly has been a great reality check for me. In my reading over the years I have become abundantly aware of the power of our "man servant" called crude oil that we use every day to make our lives easier. Three tablespoons of crude oil refined into gasoline in my rototiller does more work than I can do physically with a hoe in 8 hours. In fact, if I worked every day, all day in the garden with a hoe and a shovel, the total cumulative energy I could produce in my lifetime is equivalent to 3 barrels of crude oil. Today worldwide we extract 85 million barrels of crude oil a day. That's about 1,000 barrels a second. Or 5,500 Olympic sized swimming pools a day.

Cam using some of that amazing gasoline to make his life in the garden much easier.

As I become more aware of the precarious nature of the crude oil supply, which peaked in 2006, I pursue more and more ways to replace oil or gasoline with electricity. Electricity is something that I can make myself with my solar panels. While I now borrow my neighbor's log splitter for that 10th of my wood that doesn't want to split conveniently with an axe, I'm looking at purchasing an electric log splitter. Right now I buy distilled water to replace what's used up by batteries, but I am looking to buy a small water distiller to use extra electricity on those days when all my systems are full. When I buy distilled water someone else has purchased grid electricity to produce it. Soon I'll make it myself.

This summer we bought a small 5,000 BTU window air conditioner. During heat waves with sun everyday, we just had too much energy. I spend the day dumping excess electricity into an extra hot water tank but during a heat wave, who wants a hot bath? So now we're taking that extra electricity and running an efficient air conditioner. It keeps the bedroom cool so we can sleep. Judging by how little of our daily output it is using, I may be able to purchase additional units and cool most of the house. The beauty of constantly upgrading our systems has been to allow us to do things we never dreamed would be possible off grid.

This is the joy of living off the electricity grid. It keeps you constantly thinking about ways to make yourself more independent. This in turn means that you're producing less greenhouse gases and becoming part of the solution to climate change. This is simply a great feeling.

Michelle and I are enjoying having more and more people spend time here as guests to learn about off grid and sustainable living. It's good to have city people here to remind us how special this place is. It's easy when you live in paradise to forget just how amazing it is. It's easy to forget how much more pleasing it is to be surrounded by green on all sides, rather than by concrete and buildings and cars.

Michelle and I often have nightmares about living back in the city. For people who spent 36 years of their 50 years living in cities, this seems strange. But our psyches have acclimatized to country life very deeply and very quickly. I do not regret anything about our move to this place. I love our little 150 acres of paradise. I've picked out the place where I want my ashes sprinkled - in a cathedral of pines in the middle of the property. When my time here is done I hope my ashes will become part of this soil, part of this landscape, part of these trees that surround us. It only makes sense since I believe my soul will be here for a very, very long time after I leave this mortal plane.

Our little house off the grid is a very, very special place. I hope you are happy where you are, and if you are not, that you too can find such a place. The road may be long and winding, the path may be bumpy, but the destination will be worth the effort.

A sunny day on the front porch. The trees are green, the wood pile is stacked for next winter, the sun has charged the batteries, the hot water tanks are full, the garden is growing and life at our little house off the grid is very, very good.

Recommended Reading

Books that Have Inspired and Informed Us

Heading Home: *On Starting a New Life in a Country Place*
Lawrence Scanlan

The Pelee Project: *One Woman's Escape from Urban Madness*
Jane Christmas

Harrowsmith Magazine and later Harrowsmith Country Life Magazine
Published from 1976 until 2011.

May All Be Fed: Diet for a New World
John Robbins

Books About Peak Oil and Energy

Why Your World Is About to Get a Whole Lot Smaller
Jeff Rubin

Twilight in the Desert - *The Coming Saudi Oil Shock and the World Economy*
Matthew R. Simmons

Beyond Oil - *The View from Hubbert's Peak*
Kenneth S. Deffeyes

Books about The Financial Crisis

Empire of Debt: *The Rise of an Epic Financial Crisis*
William Bonner and Addison Wiggin

Mobs, Messiahs, and Markets: *Surviving the Public Spectacle in Finance and Politics*
William Bonner and Lila Rajiva

Bad Money: *Reckless Finance, Failed Politics, and the Global Crisis of American Capitalism*
Kevin Phillips

The Return of Depression Economics and the Crisis of 2008
Paul Krugman

Financial Armageddon: *Protecting Your Future from Four Impending Catastrophes*
Micheal J. Panzner

Books About Challenging Times

The Upside of Down: *Catastrophe, Creativity and the Renewal of Civilization*
Thomas Homer-Dixon

The Long Emergency: *Surviving the Converging Catastrophes of the Twenty-First Century*
James Howard Kunstler

Books about Personal Finance

Your Money or Your Life
Transforming your Relationship with money and achieving financial independence
Vicki Robin & Joe Dominguez with Monique Tilford

The Progress Paradox:
How Life Gets Better While People Feel Worse
Gregg Easterbrook

Books about Renewable Energy and Sustainability

The Renewable Energy Handbook
The Updated Comprehensive Guide to Renewable Energy and Independent Living
William H. Kemp

Biodiesel Basics and Beyond
A Comprehensive Guide to Production and Use for the Home and Farm
William H. Kemp

Ecoholic, *Your Guide to the Most Environmentally Friendly Information, Products, and Services*
Adria Vasil

Root Cellaring – *Natural Cold Storage of Fruits and Vegetables*
Mike & Nancy Bubel

Seed to Seed, *Seed Saving and Growing Techniques for Vegetable Gardeners*
Suzanne Ashworth

The All You Can Eat Gardening Handbook - *Easy Organic Vegetables and More Money in Your Pocket*
Cam Mather

About the Authors

Cam Mather and his wife Michelle live independently off the electricity grid using the sun and wind to power their home and their business, Aztext Press. They publish books and DVDs about renewable energy and sustainability. They have produced best-selling DVDs on organic vegetable gardening and installing a home-scale wind turbine. They have been gardening organically for 35 years and operate a market garden. They speak to large and varied groups about all aspects of sustainable living, from renewable energy to the importance of personal food independence.

For more information about our workshops visit
http://www.cammather.com/

and follow our blog at:
http://aztextpress.wordpress.com/
http://www.cammather.com/
http://aztext.com/blog/
http://motherearthnews.com/cam

Also by Cam Mather

Grow your own organic vegetables and enjoy a "One Hundred-Foot Diet!

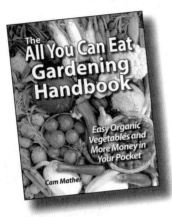

The All You Can Eat Gardening Handbook

Easy Organic Vegetables and More Money in Your Pocket

Cam Mather

260 pages 8" x 10"

ISBN 978-0-9810132-2-0

$24.95 Cdn/US

While many books make vegetable gardening look difficult with charts and checklists and talk of trace minerals and hard to find soil supplements, growing vegetables can seem intimidating. That's why *The All You Can Eat Gardening Handbook* is such a breath of fresh air. It assures readers that there's nothing to it, and encourages them to just get out there and do it. With basic tips and techniques it provides enough tools to inspire gardeners but doesn't overwhelm them.

The North American diet uses lots of fossil fuels and as we run out of the easy oil we will spend an ever-increasing percentage of our incomes on food. With many hard hit by the economic crisis, growing your own food simply makes economic sense. *The All You Can Eat Gardening Handbook* examines the health benefits of each of the vegetables and fruits listed. Sure, Grandma always told us to eat our vegetables but as adults it's nice to know about all the incredible health benefits of each item you're growing. The book also provides strategies for harvesting rainwater, watering with drip irrigation and dealing with some of the challenges our changing climate may throw at you.

Whether you have a small lot in the city, a suburban backyard or a large country property, *The All You Can Eat Gardening Handbook* is the tool you need to get motivated to start growing healthy, local, inexpensive and organic vegetables and eating the 100-Foot Diet today!

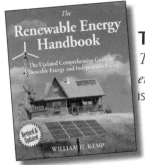

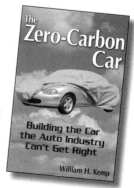

AZTEXT PRESS

DVDs from the Renewable Energy Publisher

DVDs from the Renewable Energy Publisher

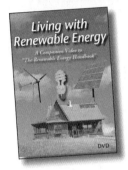

Living with Renewable Energy

This 2 hour companion to "The Renewable Energy Handbook" is a tour of several off-grid homes, including author William Kemp. It shows how to enjoy a typical North American lifestyle, powered by the sun and wind. It also includes interviews with the author about renewable energy and sustainability. ISBN 978-0-9733233-8-2

Biodiesel Basics

The companion DVD to "Biodiesel Basics and Beyond", this is a one-hour tour of author William Kemp's small scale biodiesel production facility and shows how to produce ASTM quality biodiesel from waste vegetable oil.
ISBN 978-0-9733233-7-5

Grow Your Own Vegetables

With rising fuel and food costs, this 2 hour DVD provides everything you need to turn your backyard into your own personal produce department. This program covers soil preparation, starting seeds, planting, weeding and watering, dealing with pests, and harvesting and storage of your bounty
ISBN 978-0-9733233-9-9

Home-Scale Wind Turbine Installation

This video is a step-by-step guide to putting up a home sized wind turbine using a common tubular steel tilt-up tower and winch. From evaluating your location, installing anchors, wiring, assembling the tower and using the winch to properly raise the tower, this DVD will guide and inspire your move to green energy. ISBN 978-0-9810132-0-6

AZTEXT PRESS

For more information
www.aztext.com